Analytical Chemistry

Analytical Chemistry

Ezra Flores

NY RESEARCH PRESS
New York

Published by NY Research Press
118-35 Queens Blvd., Suite 400,
Forest Hills, NY 11375, USA
www.nyresearchpress.com

Analytical Chemistry
Ezra Flores

International Standard Book Number: 978-1-63238-791-2 (Hardback)

Cataloging-in-Publication Data

Analytical chemistry / Ezra Flores.
p. cm.
Includes bibliographical references and index.
ISBN 978-1-63238-791-2
1. Chemistry, Analytic. 2. Chemistry. I. Flores, Ezra.
QD75.22 .A53 2020
543--dc23

Printed in China.

TABLE OF CONTENTS

PREFACE

This book is a culmination of my many years of practice in this field. I attribute the success of this book to my support group. I would like to thank my parents who have showered me with unconditional love and support and my peers and professors for their constant guidance.

The field of chemistry which deals with the study and utilization of instruments and methods that are used to separate, quantify and identify matter is referred to as analytical chemistry. It can be divided into classical and modern methods. The classical methods have been divided into qualitative analysis and quantitative analysis. Some common classical methods include chemical tests, flame tests, and gravimetric analysis and volumetric analysis. Instrumental methods include spectroscopy, mass spectrometry, electrochemical analysis, thermal analysis and microscopy. Analytical chemistry has applications in various fields including forensic science, clinical analysis, environmental analysis, bioanalysis, clinical analysis and materials analysis. This book provides comprehensive insights into the field of analytical chemistry. It strives to provide a fair idea about this discipline and to help develop a better understanding of the latest advances within this field. Scientists and students actively engaged in this field will find this book full of crucial and unexplored concepts.

The details of chapters are provided below for a progressive learning:

Chapter - What is Analytical Chemistry?

Analytical chemistry refers to the branch of chemistry that deals with the study and analysis, separation, identification, and quantification of matter. It is applied in various fields such as forensic science, environmental analysis, clinical analysis, medicine and chromatography. The topics elaborated in this chapter will help in gaining a better perspective about analytical chemistry and its applications.

Chapter - Techniques used in Analytical Chemistry

There are a number of techniques which are used in analytical chemistry. Some of these techniques are spectroscopy, mass spectrometry, fluorescence spectroscopy, calorimetry and chromatography. The diverse applications of these techniques have been thoroughly discussed in this chapter.

Chapter - Quantitative Analysis

The branch of chemistry which is concerned with the determination of the percentage or amount of one or more constituents of a sample is referred to as quantitative analysis. A few of the methods of quantitative analysis include titration, gravimetric analysis, Kjeldahl method and combustion analysis. This chapter closely examines these different methods associated with quantitative analysis.

Chapter - Qualitative Analysis

The branch of chemistry that focuses on the identification or grouping of elements that are present in a sample is known as qualitative analysis. It can be broadly divided into quantitative analysis of anions and quantitative analysis of cations. The topics elaborated in this chapter will help in gaining a better perspective about these aspects of quantitative analysis.

Chapter – Equilibrium Chemistry

The condition in a reversible chemical reaction in which there is no net change in the amount of products and reactants is known as chemical equilibrium. Different aspects of equilibrium chemistry include law of chemical equilibrium, equilibrium constant, etc. This chapter has been carefully written to provide an easy understanding of these aspects of equilibrium chemistry.

Ezra Flores

What is Analytical Chemistry? 1

- **Importance of Analytical Chemistry**
- **Applications of Analytical Chemistry**

Analytical chemistry refers to the branch of chemistry that deals with the study and analysis, separation, identification, and quantification of matter. It is applied in various fields such as forensic science, environmental analysis, clinical analysis, medicine and chromatography. The topics elaborated in this chapter will help in gaining a better perspective about analytical chemistry and its applications.

Analytical chemistry is the branch of chemistry that deals with the analysis of different substances.

It involves the separation, identification, and the quantification of matter. It involves the use of classical methods along with modern methods involving the use of scientific instruments.

Analytical chemistry involves the following methods:

- The process of separation isolates the required chemical species which is to be analyzed from a mixture.
- The identification of the analyte substance is achieved via the method of qualitative analysis.
- The concentration of the analyte in a given mixture can be determined with the method of quantitative analysis.

Today, the field of analytical chemistry generally involves the use of modern, sophisticated instruments. However, the principles upon which these instruments are built can be traced to more traditional techniques.

Methods used in Analytical Chemistry

The methods used to determine the identity and the quantity of the analytes in the field of analytical chemistry can be broadly divided into classical and instrumental methods.

Classical Methods

- There exist many classical methods of checking for the presence or absence of a particular compound in a given analyte. One such example is the acid test for gold.

- Another example of a classical method for qualitative analysis is the Kastle-Meyer test which employs phenolphthalein as an indicator to check for the presence of haemoglobin in the given analyte.
- Flame tests can be used to check for the presence of specific elements in an analyte by exposing it to a flame and observing the change in the color of the flame.
- Gravimetric analysis is a classical method of quantitative analysis, which can be used in analytical chemistry to determine the amount of water in a hydrate by heating it and calculating the weight of the water lost.
- One of the better known classical methods of quantitative analysis is volumetric analysis (also known as titration). In the titration method, a reactant is added to the analyte till an equivalence point is obtained.

Instrumental Methods

- Spectroscopy involves the measurement of the interaction between electromagnetic radiation and the atoms or molecules belonging to a sample.
- With the help of electric fields and magnetic fields, the method of mass spectroscopy is used to measure the ratio of the mass of the molecule to its charge.
- A common instrumental method used in the field of analytical chemistry is electrochemical analysis. In this method, the analyte is placed in an electrochemical cell and the voltage or the current flowing through it is measured.
- The interaction between the analyte and energy in the form of heat is studied in the discipline of analytical chemistry known as calorimetry. A calorimeter is an instrument that is used to measure the heat of a chemical reaction.

It can be noted that even biological measurements are made with the help of this branch of chemistry, and this field is known as bioanalytical chemistry.

Branches of Analytical Chemistry

Two sub-branches come under analytical chemistry namely quantitative analysis and qualitative analysis which can be explained as follows. These two methods form the backbone of many educational labs of analytical chemistry.

Quantitative Analysis

Quantitative Analysis is a method of determining the absolute or relative quantity regarding the concentration of one or more substances present in a sample or compound.

For example, Take a sample of an unknown solid substance. The chemists first use "qualitative" methods to identify what type of compound is present in the sample; then he adopts the quantitative analysis procedure to determine the exact amount or the quantity of the compound present in the sample.

Some of the Quantitative analysis techniques include Gravimetric Analysis and Volumetric analysis.

Qualitative Analysis

Quality means the standard or the feature of one substance. Hence Qualitative analysis method deals with the determination of the quality of a particular compound, irrespective of its quantity or concentration. In simpler words, the qualitative analysis does not measure the amount of the substance but measures the quality of that material. One of the best examples of this type of method is the observation of a chemical reaction, whether there will be a change in color or not.

The qualitative analysis method can be measured in different ways such as Chemical tests, flame tests, etc.

Importance of Analytical Chemistry

Analytical chemistry is the branch which is taught in almost all schools and colleges. But the applications of it are made in pharmaceutical industries, food factories, chemical industries, agricultural industries and in scientific laboratories. The tools used for this purpose are quite expensive which one cannot afford at home.

Importance in Pharmacy

Pharmacy is one of the subjects where analytical chemistry is studied vastly. It is used for purposes like checking the quality of inorganic compounds, organic compounds, drugs and other useful chemicals.

It is mostly used for purposes like to determine shelf life, dissolution studies and contaminants.

To Determine the Shelf-life of a Compound

Self-life is the period till which a compound will be fit for use. For medicines, it is the time required for 90% of the drug ingredient to stay active after the date of manufacture. Hence, we can find an expiry date marked on all the medicine packages. This means a medicine cannot be sold if its expiry date is elapsed.

Further, one can notice that in a pharmacy some medicines are stored in the refrigerator while some are stored in dark containers. These steps are taken to maintain the shelf-life by preventing the rapid degradation of the medication.

To Determine Adulterants

During the manufacture of drugs, there are many chemical intermediates and reagents used. So there are many chances that impurities like heavy metals, dirt, etc. can get incorporated into the final preparation. Analytical chemistry is used to see if the formulation is within limits of contamination. If the contamination is high, then that batch of preparation is unfit for consumption.

Drug Dissolution Studies

When a medical formulation is made in the form of a tablet, ointment, capsule etc. It is tested for its ability to release the medicine from within when put in a suitable medium. The dissolution studies are done to see if the medicine is released completely from the tablet. Also, the time taken for release. Both these factors are essential because when a tablet is swallowed, it stays in the stomach only for an hour. So if it has to release the medicine, it has to do it within one hour. Also, medicine shows its effects at a suitable dose and not sub-doses. So the tablet has to release the entire content within the said time. Hence dissolution studies are done by using analytical chemistry to test the release of drug from a formulation.

Importance in the Environment

Similarly, in environmental studies, it is widely used for the study of air condition (pollution), water and soil contents in a particular locale.

In Medical Technology and Research

This is also a field which is highly reliant on analytical chemistry. In health care, it finds its use in diagnosis. Simple tests like serum cholesterol, urine ketones, blood glucose level rely on analytical chemistry.

For example, blood glucose is estimated by many methods. But one of the common technique is to convert the glucose into a colored substance in a beaker and analyze it by titration with a reacting mixture.

Similarly, serum cholesterol is centrifuged and then a coloring substance is added. The superficial layer having cholesterol is (which is already colored) is taken into a cuvette and quantified by spectrophotometry. Thus we regularly use analytical chemistry and its principles for a various diagnostic test in health care.

So what are the methods used for the purpose? There are many methods of analysis but some of them are routinely used as they are inexpensive, less time taking and versatile (suitable for many).

Some of the methods include titrations, especially acid-base, chromatography, spectroscopy, etc.

Applications of Analytical Chemistry

Some important applications of this branch of chemistry are listed below:

- The shelf lives of many medicines are determined with the help of analytical chemistry.
- It is used to check for the presence of adulterants in drugs.
- Soil can be tested to check for appropriate concentrations of minerals and nutrients that are necessary for plant growth.
- It is employed in the process of chromatography where the blood samples of a person are classified.

- The concentration of the pesticide residues and the contaminants in a given food sample can also be determined via analytical chemistry.
- It also has many important applications in medicine, with its use in the testing of cholesterol and glucose levels in a blood sample.
- Analytical chemistry is an integral part of forensic science, clinical analysis, and even environmental analysis.

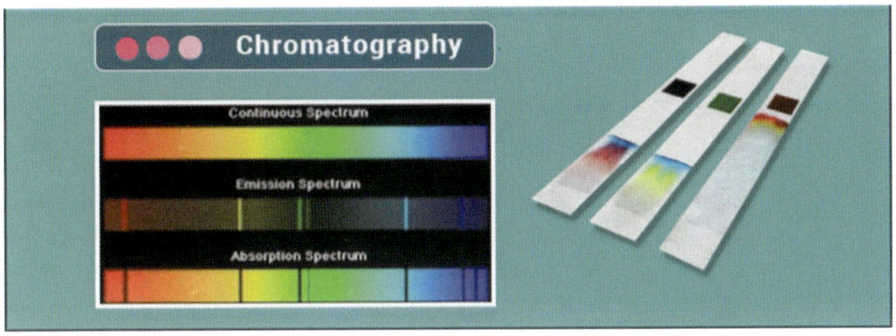

Applications of analytical chemistry in chromatography.

Applications in Agriculture

It is used for 3 aspects in agriculture like in:

- Soil testing: This is done to study the inorganic and organic minerals in the soil. This tells us how far the soil is suitable for the growth of a particular crop. It also gives an idea of the microbial environment, fertility and pesticide residues in the soil.
- Water testing: Not all water is suitable for the growth of crops. Some of the water is very hard that it deposits lime on the fertile soil making it unfit for further cultivation. By analytical chemistry, one can analyze the water for its constituents and make suitable changes to it.
- Harvested crop testing. Once the crop is harvested, the product is tested for its quality in terms of the pesticide residues, any contaminants, etc. If the plant material has a high amount of pesticide or other waste, it is rejected for human use.

Techniques used in Analytical Chemistry

2

- Spectroscopy
- Mass Spectrometry
- Fluorescence Spectroscopy
- Electroanalytical Methods
- Chromatography
- Calorimetry
- Surface Analysis

There are a number of techniques which are used in analytical chemistry. Some of these techniques are spectroscopy, mass spectrometry, fluorescence spectroscopy, calorimetry and chromatography. The diverse applications of these techniques have been thoroughly discussed in this chapter.

Spectroscopy

Spectroscopy is study of the absorption and emission of light and other radiation by matter, as related to the dependence of these processes on the wavelength of the radiation. More recently, the definition has been expanded to include the study of the interactions between particles such as electrons, protons, and ions, as well as their interaction with other particles as a function of their collision energy. Spectroscopic analysis has been crucial in the development of the most fundamental theories in physics, including quantum mechanics, the special and general theories of relativity, and quantum electrodynamics. Spectroscopy, as applied to high-energy collisions, has been a key tool in developing scientific understanding not only of the electromagnetic force but also of the strong and weak nuclear forces.

Spectroscopic techniques have been applied in virtually all technical fields of science and technology. Radio-frequency spectroscopy of nuclei in a magnetic field has been employed in a medical technique called magnetic resonance imaging (MRI) to visualize the internal soft tissue of the body with unprecedented resolution. Microwave spectroscopy was used to discover the so-called three-degree blackbody radiation, the remnant of the big bang (i.e., the primeval explosion) from which the universe is thought to have originated. The internal structure of the proton and neutron and the state of the early universe up to the first thousandth of a second of its existence are being unraveled with spectroscopic techniques using high-energy particle accelerators. The constituents of distant stars, intergalactic molecules, and even the primordial abundance of the elements before the formation of the first stars can be determined by optical, radio, and X-ray spectroscopy. Optical spectroscopy is used routinely to identify the chemical composition of matter and to determine its physical structure.

Spectroscopic techniques are extremely sensitive. Single atoms and even different isotopes of the same atom can be detected among 10 or more atoms of a different species. (Isotopes are all atoms of an element that have unequal mass but the same atomic number. Isotopes of the same element are virtually identical chemically.) Trace amounts of pollutants or contaminants are often detected most effectively by spectroscopic techniques. Certain types of microwave, optical, and gamma-ray spectroscopy are capable of measuring infinitesimal frequency shifts in narrow spectroscopic lines. Frequency shifts as small as one part in 10 of the frequency being measured can be observed with ultrahigh resolution laser techniques. Because of this sensitivity, the most accurate physical measurements have been frequency measurements.

Spectroscopy now covers a sizable fraction of the electromagnetic spectrum. The table summarizes the electromagnetic spectrum over a frequency range of 16 orders of magnitude. Spectroscopic techniques are not confined to electromagnetic radiation, however. Because the energy E of a photon (a quantum of light) is related to its frequency ν by the relation $E = h\nu$, where h is Planck's constant, spectroscopy is actually the measure of the interaction of photons with matter as a function of the photon energy. In instances where the probe particle is not a photon, spectroscopy refers to the measurement of how the particle interacts with the test particle or material as a function of the energy of the probe particle.

An example of particle spectroscopy is a surface analysis technique known as electron energy loss spectroscopy (EELS) that measures the energy lost when low-energy electrons (typically 5–10 electron volts) collide with a surface. Occasionally, the colliding electron loses energy by exciting the surface; by measuring the electron's energy loss, vibrational excitations associated with the surface can be measured. On the other end of the energy spectrum, if an electron collides with another particle at exceedingly high energies, a wealth of subatomic particles is produced. Most of what is known in particle physics (the study of subatomic particles) has been gained by analyzing the total particle production or the production of certain particles as a function of the incident energies of electrons and protons.

Mass Spectrometry

Mass spectrometry (MS) is an analytical technique that measures the mass-to-charge ratio of ions. The results are typically presented as a mass spectrum, a plot of intensity as a function of the mass-to-charge ratio. Mass spectrometry is used in many different fields and is applied to pure samples as well as complex mixtures.

A mass spectrum is a plot of the ion signal as a function of the mass-to-charge ratio. These spectra are used to determine the elemental or isotopic signature of a sample, the masses of particles and of molecules, and to elucidate the chemical identity or structure of molecules and other chemical compounds.

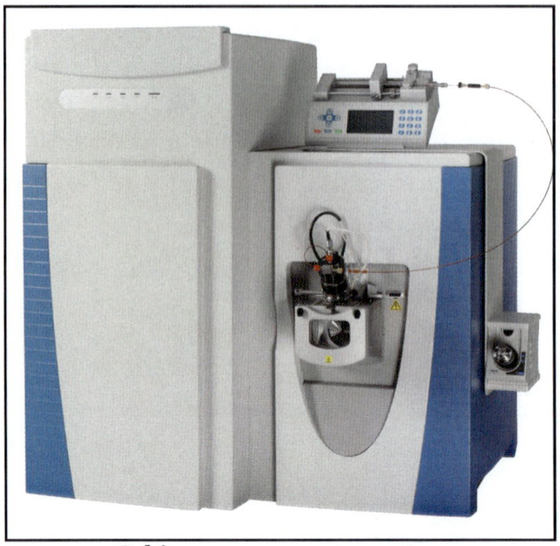

Orbitrap mass spectrometer.

In a typical MS procedure, a sample, which may be solid, liquid, or gaseous, is ionized, for example by bombarding it with electrons. This may cause some of the sample's molecules to break into charged fragments or simply become charged without fragmenting. These ions are then separated according to their mass-to-charge ratio, for example by accelerating them and subjecting them to an electric or magnetic field: ions of the same mass-to-charge ratio will undergo the same amount of deflection. The ions are detected by a mechanism capable of detecting charged particles, such as an electron multiplier. Results are displayed as spectra of the signal intensity of detected ions as a function of the mass-to-charge ratio. The atoms or molecules in the sample can be identified by correlating known masses (e.g. an entire molecule) to the identified masses or through a characteristic fragmentation pattern.

Parts of a Mass Spectrometer

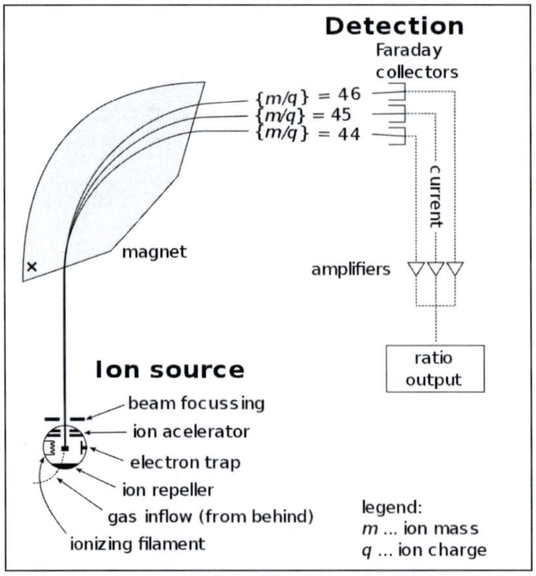

Schematics of a simple mass spectrometer with sector type mass analyzer. This one is for the measurement of carbon dioxide isotope ratios (IRMS) as in the carbon-13 urea breath test.

A mass spectrometer consists of three components: an ion source, a mass analyzer, and a detector. The *ionizer* converts a portion of the sample into ions. There is a wide variety of ionization techniques, depending on the phase (solid, liquid, gas) of the sample and the efficiency of various ionization mechanisms for the unknown species. An extraction system removes ions from the sample, which are then targeted through the mass analyzer and into the *detector*. The differences in masses of the fragments allows the mass analyzer to sort the ions by their mass-to-charge ratio. The detector measures the value of an indicator quantity and thus provides data for calculating the abundances of each ion present. Some detectors also give spatial information, e.g., a multichannel plate.

Theoretical Example

The following example describes the operation of a spectrometer mass analyzer, which is of the sector type. Consider a sample of sodium chloride (table salt). In the ion source, the sample is vaporized (turned into gas) and ionized (transformed into electrically charged particles) into sodium (Na^+) and chloride (Cl^-) ions. Sodium atoms and ions are monoisotopic, with a mass of about 23 u. Chloride atoms and ions come in two isotopes with masses of approximately 35 u (at a natural abundance of about 75 percent) and approximately 37 u (at a natural abundance of about 25 percent). The analyzer part of the spectrometer contains electric and magnetic fields, which exert forces on ions traveling through these fields. The speed of a charged particle may be increased or decreased while passing through the electric field, and its direction may be altered by the magnetic field. The magnitude of the deflection of the moving ion's trajectory depends on its mass-to-charge ratio. Lighter ions get deflected by the magnetic force more than heavier ions (based on Newton's second law of motion, $F = ma$). The streams of sorted ions pass from the analyzer to the detector, which records the relative abundance of each ion type. This information is used to determine the chemical element composition of the original sample (i.e. that both sodium and chlorine are present in the sample) and the isotopic composition of its constituents (the ratio of ^{35}Cl to ^{37}Cl).

Creating Ions

Surface ionization source at the Argonne National Laboratory linear accelerator.

The ion source is the part of the mass spectrometer that ionizes the material under analysis (the analyte). The ions are then transported by magnetic or electric fields to the mass analyzer.

Techniques for ionization have been key to determining what types of samples can be analyzed by mass spectrometry. Electron ionization and chemical ionization are used for gases and vapors.

In chemical ionization sources, the analyte is ionized by chemical ion-molecule reactions during collisions in the source. Two techniques often used with liquid and solid biological samples include electrospray ionization (invented by John Fenn) and matrix-assisted laser desorption/ionization (MALDI, initially developed as a similar technique "Soft Laser Desorption (SLD)" by K. Tanaka for which a Nobel Prize was awarded and as MALDI by M. Karas and F. Hillenkamp).

Hard Ionization and Soft Ionization

Quadrupole mass spectrometer and electrospray ion source used for Fenn's early work.

In mass spectrometry, ionization refers to the production of gas phase ions suitable for resolution in the mass analyser or mass filter. Ionization occurs in the ion source. There are several ion sources available; each has advantages and disadvantages for particular applications. For example, electron ionization (EI) gives a high degree of fragmentation, yielding highly detailed mass spectra which when skilfully analysed can provide important information for structural elucidation/characterisation and facilitate identification of unknown compounds by comparison to mass spectral libraries obtained under identical operating conditions. However, EI is not suitable for coupling to HPLC, i.e. LC-MS, since at atmospheric pressure, the filaments used to generate electrons burn out rapidly. Thus EI is coupled predominantly with GC, i.e. GC-MS, where the entire system is under high vacuum.

Hard ionization techniques are processes which impart high quantities of residual energy in the subject molecule invoking large degrees of fragmentation (i.e. the systematic rupturing of bonds acts to remove the excess energy, restoring stability to the resulting ion). Resultant ions tend to have m/z lower than the molecular mass (other than in the case of proton transfer and not including isotope peaks). The most common example of hard ionization is electron ionization (EI).

Soft ionization refers to the processes which impart little residual energy onto the subject molecule and as such result in little fragmentation. Examples include fast atom bombardment (FAB), chemical ionization (CI), atmospheric-pressure chemical ionization (APCI), electrospray ionization (ESI), and matrix-assisted laser desorption/ionization (MALDI).

Inductively Coupled Plasma

Inductively coupled plasma (ICP) sources are used primarily for cation analysis of a wide array of sample types. In this source, a plasma that is electrically neutral overall, but that has had a substantial fraction of its atoms ionized by high temperature, is used to atomize introduced sample molecules and to further strip the outer electrons from those atoms. The plasma is usually generated from argon gas, since the first ionization energy of argon atoms is higher than the first of any other elements except He, O, F and Ne, but lower than the second ionization energy of all except the most electropositive metals. The heating is achieved by a radio-frequency current passed through a coil surrounding the plasma.

Inductively coupled plasma ion source.

Photoionization Mass Spectrometry

Photoionization can be used in experiments which seek to use mass spectrometry as a means of resolving chemical kinetics mechanisms and isomeric product branching. In such instances a high energy photon, either X-ray or uv, is used to dissociate stable gaseous molecules in a carrier gas of He or Ar. In instances where a synchrotron light source is utilized, a tuneable photon energy can be utilized to acquire a photoionization efficiency curve which can be used in conjunction with the charge ratio m/z to fingerprint molecular and ionic species.

Other Ionization Techniques

Others include glow discharge, field desorption (FD), fast atom bombardment (FAB), thermospray, desorption/ionization on silicon (DIOS), Direct Analysis in Real Time (DART), atmospheric pressure chemical ionization (APCI), secondary ion mass spectrometry (SIMS), spark ionization and thermal ionization (TIMS).

Mass Selection

Mass analyzers separate the ions according to their mass-to-charge ratio. The following two laws govern the dynamics of charged particles in electric and magnetic fields in vacuum:

$\mathbf{F} = Q(\mathbf{E} + \mathbf{v} \times \mathbf{B})$ (Lorentz force law);

$\mathbf{F} = m\mathbf{a}$ (Newton's second law of motion in non-relativistic case, i.e. valid only at ion velocity much lower than the speed of light).

Here $\mathbf{F}$ is the force applied to the ion, m is the mass of the ion, $\mathbf{a}$ is the acceleration, Q is the ion charge, $\mathbf{E}$ is the electric field, and $\mathbf{v} \times \mathbf{B}$ is the vector cross product of the ion velocity and the magnetic field.

Equating the expressions for the force applied to the ion yields:

$$(m / Q)\mathbf{a} = \mathbf{E} + \mathbf{v} \times \mathbf{B}$$

This differential equation is the classic equation of motion for charged particles. Together with the particle's initial conditions, it completely determines the particle's motion in space and time in terms of m/Q. Thus mass spectrometers could be thought of as "mass-to-charge spectrometers". When presenting data, it is common to use the (officially) dimensionless m/z, where z is the number of elementary charges (e) on the ion (z=Q/e). This quantity, although it is informally called the mass-to-charge ratio, more accurately speaking represents the ratio of the mass number and the charge number, z.

There are many types of mass analyzers, using either static or dynamic fields, and magnetic or electric fields, but all operate according to the differential equation. Each analyzer type has its strengths and weaknesses. Many mass spectrometers use two or more mass analyzers for tandem mass spectrometry (MS/MS).

There are several important analyser characteristics. The mass resolving power is the measure of the ability to distinguish two peaks of slightly different m/z. The mass accuracy is the ratio of the m/z measurement error to the true m/z. Mass accuracy is usually measured in ppm or milli mass units. The mass range is the range of m/z amenable to analysis by a given analyzer. The linear dynamic range is the range over which ion signal is linear with analyte concentration. Speed refers to the time frame of the experiment and ultimately is used to determine the number of spectra per unit time that can be generated.

Sector Instruments

ThermoQuest AvantGarde sector mass spectrometer.

A sector field mass analyzer uses a static electric and magnetic field to affect the path and velocity of the charged particles in some way. The sector instruments bend the trajectories of the ions as they pass through the mass analyzer, according to their mass-to-charge ratios, deflecting the more charged and faster-moving, lighter ions more. The analyzer can be used to select a narrow range of m/z or to scan through a range of m/z to catalog the ions present.

Time of Flight

The time of flight (TOF) analyzer uses an electric field to accelerate the ions through the same potential, and then measures the time they take to reach the detector. If the particles all have the same charge, their kinetic energies will be identical, and their velocities will depend only on their masses. Ions with a lower mass will reach the detector first. However, in reality, even particles with the same m/z can arrive at different times at the detector, because they have different initial velocities. The initial velocity is often not dependent on the mass of the ion TOF-MS, and will turn into a difference in the final velocity. Because of this, ions with the same m/z ratio will reach the detector at a variety of times, which broadens the peaks shown on the count vs m/z plot, but will generally not change the central location of the peaks, since the average starting velocity of ions relative to the other analyzed ions is generally centered at zero. To fix this problem, time-lag focusing/delayed extraction has been coupled with TOF-MS.

Quadrupole Mass Filter

Quadrupole mass analyzers use oscillating electrical fields to selectively stabilize or destabilize the paths of ions passing through a radio frequency (RF) quadrupole field created between 4 parallel rods. Only the ions in a certain range of mass/charge ratio are passed through the system at any time, but changes to the potentials on the rods allow a wide range of m/z values to be swept rapidly, either continuously or in a succession of discrete hops. A quadrupole mass analyzer acts as a mass-selective filter and is closely related to the quadrupole ion trap, particularly the linear quadrupole ion trap except that it is designed to pass the untrapped ions rather than collect the trapped ones, and is for that reason referred to as a transmission quadrupole. A magnetically enhanced quadrupole mass analyzer includes the addition of a magnetic field, either applied axially or transversely. This novel type of instrument leads to an additional performance enhancement in terms of resolution and sensitivity depending upon the magnitude and orientation of the applied magnetic field. A common variation of the transmission quadrupole is the triple quadrupole mass spectrometer. The “triple quad” has three consecutive quadrupole stages, the first acting as a mass filter to transmit a particular incoming ion to the second quadrupole, a collision chamber, wherein that ion can be broken into fragments. The third quadrupole also acts as a mass filter, to transmit a particular fragment ion to the detector. If a quadrupole is made to rapidly and repetitively cycle through a range of mass filter settings, full spectra can be reported. Likewise, a triple quad can be made to perform various scan types characteristic of tandem mass spectrometry.

Ion Traps

Three-dimensional Quadrupole Ion Trap

The quadrupole ion trap works on the same physical principles as the quadrupole mass analyzer, but the ions are trapped and sequentially ejected. Ions are trapped in a mainly quadrupole RF field, in a space defined by a ring electrode (usually connected to the main RF potential) between two endcap electrodes (typically connected to DC or auxiliary AC potentials). The sample is ionized either internally (e.g. with an electron or laser beam), or externally, in which case the ions are often introduced through an aperture in an endcap electrode.

There are many mass/charge separation and isolation methods but the most commonly used is the mass instability mode in which the RF potential is ramped so that the orbit of ions with a mass $a > b$ are stable while ions with mass b become unstable and are ejected on the z-axis onto a detector. There are also non-destructive analysis methods.

Ions may also be ejected by the resonance excitation method, whereby a supplemental oscillatory excitation voltage is applied to the endcap electrodes, and the trapping voltage amplitude and excitation voltage frequency is varied to bring ions into a resonance condition in order of their mass/charge ratio.

Cylindrical Ion Trap

The cylindrical ion trap mass spectrometer (CIT) is a derivative of the quadrupole ion trap where the electrodes are formed from flat rings rather than hyperbolic shaped electrodes. The architecture lends itself well to miniaturization because as the size of a trap is reduced, the shape of the electric field near the center of the trap, the region where the ions are trapped, forms a shape similar to that of a hyperbolic trap.

Linear Quadrupole Ion Trap

A linear quadrupole ion trap is similar to a quadrupole ion trap, but it traps ions in a two dimensional quadrupole field, instead of a three-dimensional quadrupole field as in a 3D quadrupole ion trap. Thermo Fisher's LTQ ("linear trap quadrupole") is an example of the linear ion trap.

A toroidal ion trap can be visualized as a linear quadrupole curved around and connected at the ends or as a cross section of a 3D ion trap rotated on edge to form the toroid, donut shaped trap. The trap can store large volumes of ions by distributing them throughout the ring-like trap structure. This toroidal shaped trap is a configuration that allows the increased miniaturization of an ion trap mass analyzer. Additionally all ions are stored in the same trapping field and ejected together simplifying detection that can be complicated with array configurations due to variations in detector alignment and machining of the arrays.

As with the toroidal trap, linear traps and 3D quadrupole ion traps are the most commonly miniaturized mass analyzers due to their high sensitivity, tolerance for mTorr pressure, and capabilities for single analyzer tandem mass spectrometry (e.g. product ion scans).

Orbitrap

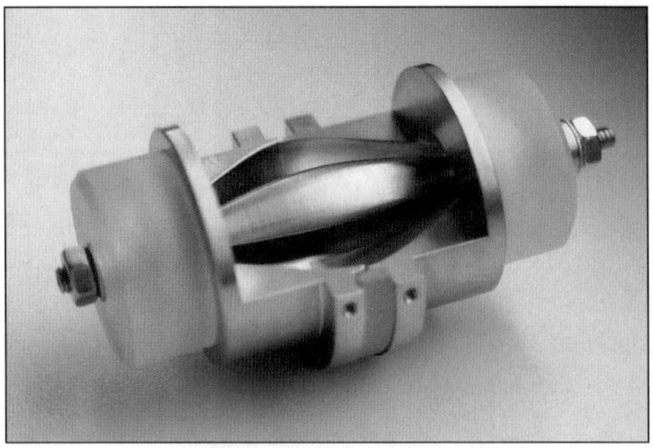

Orbitrap mass analyzer.

Orbitrap instruments are similar to Fourier transform ion cyclotron resonance mass spectrometers. Ions are electrostatically trapped in an orbit around a central, spindle shaped electrode. The electrode confines the ions so that they both orbit around the central electrode and oscillate back and forth along the central electrode's long axis. This oscillation generates an image current in the detector plates which is recorded by the instrument. The frequencies of these image currents depend on the mass-to-charge ratios of the ions. Mass spectra are obtained by Fourier transformation of the recorded image currents.

Orbitraps have a high mass accuracy, high sensitivity and a good dynamic range.

Fourier Transform Ion Cyclotron Resonance

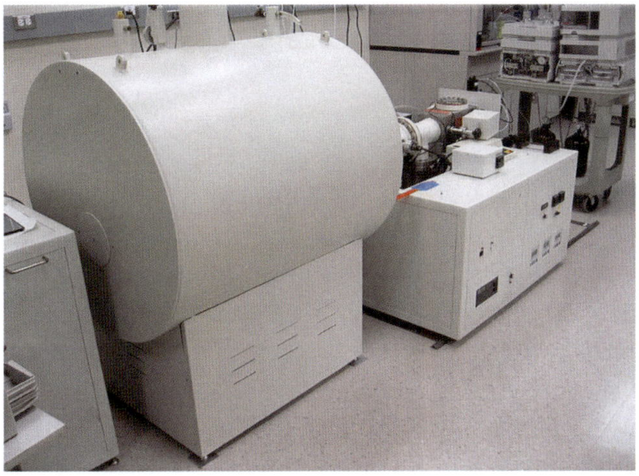

Fourier transform ion cyclotron resonance mass spectrometer.

Fourier transform mass spectrometry (FTMS), or more precisely Fourier transform ion cyclotron resonance MS, measures mass by detecting the image current produced by ions cyclotroning in the presence of a magnetic field. Instead of measuring the deflection of ions with a detector such as an electron multiplier, the ions are injected into a Penning trap (a static electric/magnetic ion trap) where they effectively form part of a circuit. Detectors at fixed positions in space measure the electrical signal of ions which pass near them over time, producing a periodic signal. Since the frequency of an ion's cycling is determined by its mass-to-charge ratio, this can be deconvoluted by performing a Fourier transform on the signal. FTMS has the advantage of high sensitivity (since each ion is "counted" more than once) and much higher resolution and thus precision.

Ion cyclotron resonance (ICR) is an older mass analysis technique similar to FTMS except that ions are detected with a traditional detector. Ions trapped in a Penning trap are excited by an RF electric field until they impact the wall of the trap, where the detector is located. Ions of different mass are resolved according to impact time.

Detectors

The final element of the mass spectrometer is the detector. The detector records either the charge induced or the current produced when an ion passes by or hits a surface. In a scanning instrument, the signal produced in the detector during the course of the scan versus where the instrument is in the scan (at what m/Q) will produce a mass spectrum, a record of ions as a function of m/Q.

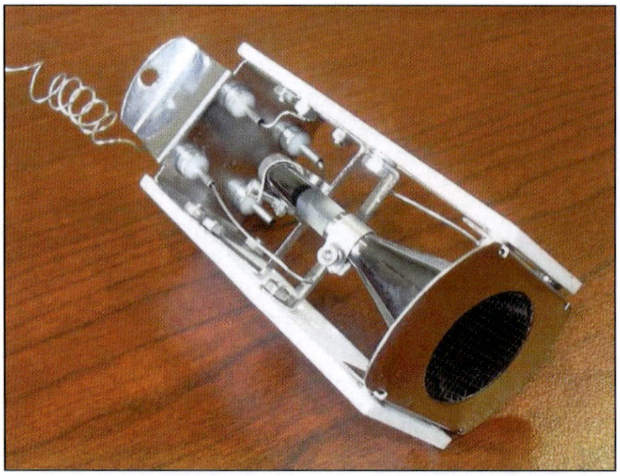

A continuous dynode particle multiplier detector.

Typically, some type of electron multiplier is used, though other detectors including Faraday cups and ion-to-photon detectors are also used. Because the number of ions leaving the mass analyzer at a particular instant is typically quite small, considerable amplification is often necessary to get a signal. Microchannel plate detectors are commonly used in modern commercial instruments. In FTMS and Orbitraps, the detector consists of a pair of metal surfaces within the mass analyzer/ion trap region which the ions only pass near as they oscillate. No direct current is produced, only a weak AC image current is produced in a circuit between the electrodes. Other inductive detectors have also been used.

Tandem Mass Spectrometry

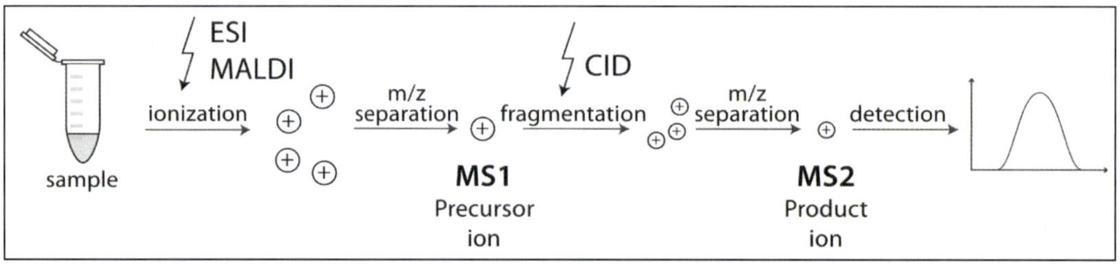

Tandem mass spectrometry for biological molecules using ESI or MALDI.

A tandem mass spectrometer is one capable of multiple rounds of mass spectrometry, usually separated by some form of molecule fragmentation. For example, one mass analyzer can isolate one peptide from many entering a mass spectrometer. A second mass analyzer then stabilizes the peptide ions while they collide with a gas, causing them to fragment by collision-induced dissociation (CID). A third mass analyzer then sorts the fragments produced from the peptides. Tandem MS can also be done in a single mass analyzer over time, as in a quadrupole ion trap. There are various methods for fragmenting molecules for tandem MS, including collision-induced dissociation (CID), electron capture dissociation (ECD), electron transfer dissociation (ETD), infrared multiphoton dissociation (IRMPD), blackbody infrared radiative dissociation (BIRD), electron-detachment dissociation (EDD) and surface-induced dissociation (SID). An important application using tandem mass spectrometry is in protein identification.

Tandem mass spectrometry enables a variety of experimental sequences. Many commercial mass spectrometers are designed to expedite the execution of such routine sequences as selected reaction monitoring (SRM) and precursor ion scanning. In SRM, the first analyzer allows only a single mass through and the second analyzer monitors for multiple user-defined fragment ions. SRM is most often used with scanning instruments where the second mass analysis event is duty cycle limited. These experiments are used to increase specificity of detection of known molecules, notably in pharmacokinetic studies. Precursor ion scanning refers to monitoring for a specific loss from the precursor ion. The first and second mass analyzers scan across the spectrum as partitioned by a user-defined *m/z* value. This experiment is used to detect specific motifs within unknown molecules.

Another type of tandem mass spectrometry used for radiocarbon dating is accelerator mass spectrometry (AMS), which uses very high voltages, usually in the mega-volt range, to accelerate negative ions into a type of tandem mass spectrometer.

Common Mass Spectrometer Configurations and Techniques

When a specific combination of source, analyzer, and detector becomes conventional in practice, a compound acronym may arise to designate it succinctly. One example is MALDI-TOF, which refers to a combination of a matrix-assisted laser desorption/ionization source with a time-of-flight mass analyzer. Other examples include inductively coupled plasma-mass spectrometry (ICP-MS), accelerator mass spectrometry (AMS), thermal ionization-mass spectrometry (TIMS) and spark source mass spectrometry (SSMS).

Certain applications of mass spectrometry have developed monikers that although strictly speaking would seem to refer to a broad application, in practice have come instead to connote a specific or a limited number of instrument configurations. An example of this is isotope ratio mass spectrometry (IRMS), which refers in practice to the use of a limited number of sector based mass analyzers; this name is used to refer to both the application and the instrument used for the application.

Separation Techniques Combined with Mass Spectrometry

An important enhancement to the mass resolving and mass determining capabilities of mass spectrometry is using it in tandem with chromatographic and other separation techniques.

Gas Chromatography

A gas chromatograph (right) directly coupled to a mass spectrometer (left).

A common combination is gas chromatography-mass spectrometry (GC/MS or GC-MS). In this technique, a gas chromatograph is used to separate different compounds. This stream of separated compounds is fed online into the ion source, a metallic filament to which voltage is applied. This filament emits electrons which ionize the compounds. The ions can then further fragment, yielding predictable patterns. Intact ions and fragments pass into the mass spectrometer's analyzer and are eventually detected.

Liquid Chromatography

Scientist performing liquid chromatography–mass spectrometry.

Similar to gas chromatography MS (GC-MS), liquid chromatography-mass spectrometry (LC/MS or LC-MS) separates compounds chromatographically before they are introduced to the ion source and mass spectrometer. It differs from GC-MS in that the mobile phase is liquid, usually a mixture of water and organic solvents, instead of gas. Most commonly, an electrospray ionization source is used in LC-MS. Other popular and commercially available LC-MS ion sources are atmospheric pressure chemical ionization and atmospheric pressure photoionization. There are also some newly developed ionization techniques like laser spray.

Capillary Electrophoresis-mass Spectrometry

Capillary electrophoresis–mass spectrometry (CE-MS) is a technique that combines the liquid separation process of capillary electrophoresis with mass spectrometry. CE-MS is typically coupled to electrospray ionization.

Ion Mobility

Ion mobility spectrometry-mass spectrometry (IMS/MS or IMMS) is a technique where ions are first separated by drift time through some neutral gas under an applied electrical potential gradient before being introduced into a mass spectrometer. Drift time is a measure of the radius relative to the charge of the ion. The duty cycle of IMS (the time over which the experiment takes place) is longer than most mass spectrometric techniques, such that the mass spectrometer can sample along the course of the IMS separation. This produces data about the IMS separation and the mass-to-charge ratio of the ions in a manner similar to LC-MS.

The duty cycle of IMS is short relative to liquid chromatography or gas chromatography separations and can thus be coupled to such techniques, producing triple modalities such as LC/IMS/MS.

Data and Analysis

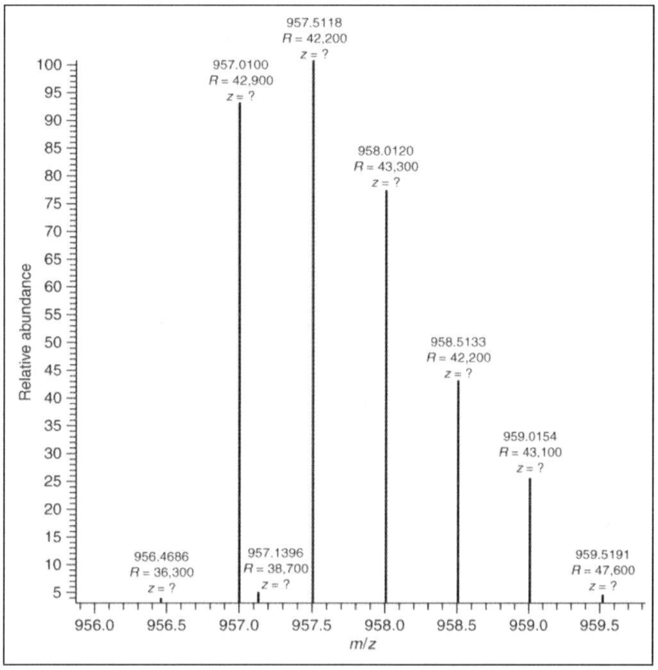

Data Representations

Mass spectrometry produces various types of data. The most common data representation is the mass spectrum.

Certain types of mass spectrometry data are best represented as a mass chromatogram. Types of chromatograms include selected ion monitoring (SIM), total ion current (TIC), and selected reaction monitoring (SRM), among many others.

Other types of mass spectrometry data are well represented as a three-dimensional contour map. In this form, the mass-to-charge, m/z is on the x-axis, intensity the y-axis, and an additional experimental parameter, such as time, is recorded on the z-axis.

Data Analysis

Mass spectrometry data analysis is specific to the type of experiment producing the data. General subdivisions of data are fundamental to understanding any data.

Many mass spectrometers work in either negative ion mode or positive ion mode. It is very important to know whether the observed ions are negatively or positively charged. This is often important in determining the neutral mass but it also indicates something about the nature of the molecules.

Different types of ion source result in different arrays of fragments produced from the original molecules. An electron ionization source produces many fragments and mostly single-charged (1-) radicals (odd number of electrons), whereas an electrospray source usually produces non-radical quasimolecular ions that are frequently multiply charged. Tandem mass spectrometry purposely produces fragment ions post-source and can drastically change the sort of data achieved by an experiment.

Knowledge of the origin of a sample can provide insight into the component molecules of the sample and their fragmentations. A sample from a synthesis/manufacturing process will probably contain impurities chemically related to the target component. A crudely prepared biological sample will probably contain a certain amount of salt, which may form adducts with the analyte molecules in certain analyses.

Results can also depend heavily on sample preparation and how it was run/introduced. An important example is the issue of which matrix is used for MALDI spotting, since much of the energetics of the desorption/ionization event is controlled by the matrix rather than the laser power. Sometimes samples are spiked with sodium or another ion-carrying species to produce adducts rather than a protonated species.

Mass spectrometry can measure molar mass, molecular structure, and sample purity. Each of these questions requires a different experimental procedure; therefore, adequate definition of the experimental goal is a prerequisite for collecting the proper data and successfully interpreting it.

Interpretation of Mass Spectra

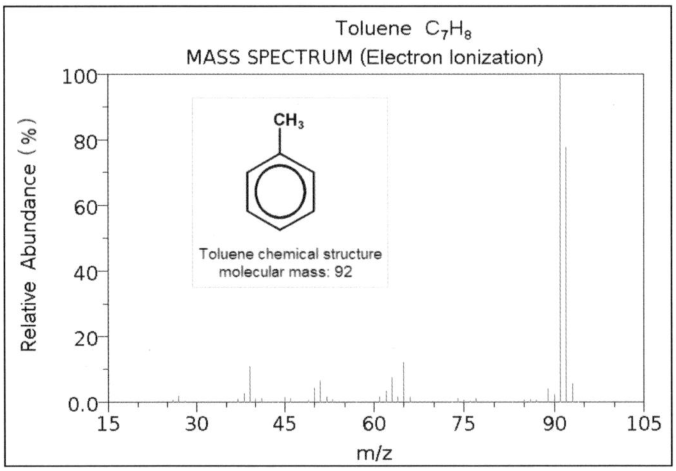

Toluene electron ionization mass spectrum.

Since the precise structure or peptide sequence of a molecule is deciphered through the set of fragment masses, the interpretation of mass spectra requires combined use of various techniques. Usually the first strategy for identifying an unknown compound is to compare its experimental mass spectrum against a library of mass spectra. If no matches result from the search, then manual interpretation or software assisted interpretation of mass spectra must be performed. Computer simulation of ionization and fragmentation processes occurring in mass spectrometer is the primary tool for assigning structure or peptide sequence to a molecule. An *a priori* structural information is fragmented *in silico* and the resulting pattern is compared with observed spectrum. Such simulation is often supported by a fragmentation library that contains published patterns of known decomposition reactions. Software taking advantage of this idea has been developed for both small molecules and proteins.

Analysis of mass spectra can also be spectra with accurate mass. A mass-to-charge ratio value (m/z) with only integer precision can represent an immense number of theoretically possible ion structures; however, more precise mass figures significantly reduce the number of candidate molecular formulas. A computer algorithm called formula generator calculates all molecular formulas that theoretically fit a given mass with specified tolerance.

A recent technique for structure elucidation in mass spectrometry, called precursor ion fingerprinting, identifies individual pieces of structural information by conducting a search of the tandem spectra of the molecule under investigation against a library of the product-ion spectra of structurally characterized precursor ions.

Applications

NOAA Particle Analysis by Laser Mass Spectrometry aerosol mass spectrometer aboard a NASA WB-57 high-altitude research aircraft.

Mass spectrometry has both qualitative and quantitative uses. These include identifying unknown compounds, determining the isotopic composition of elements in a molecule, and determining the structure of a compound by observing its fragmentation. Other uses include quantifying the amount of a compound in a sample or studying the fundamentals of gas phase ion chemistry (the chemistry of ions and neutrals in a vacuum). MS is now commonly used in analytical laboratories that study physical, chemical, or biological properties of a great variety of compounds.

As an analytical technique it possesses distinct advantages such as: Increased sensitivity over most other analytical techniques because the analyzer, as a mass-charge filter, reduces background interference, excellent specificity from characteristic fragmentation patterns to identify unknowns or confirm the presence of suspected compounds, information about molecular weight, Information about the isotopic abundance of elements, temporally resolved chemical data.

A few of the disadvantages of the method is that it often fails to distinguish between optical and geometrical isomers and the positions of substituent in o-, m- and p- positions in an aromatic ring. Also, its scope is limited in identifying hydrocarbons that produce similar fragmented ions.

Isotope Ratio MS: Isotope Dating and Tracing

Mass spectrometer.

Mass spectrometry is also used to determine the isotopic composition of elements within a sample. Differences in mass among isotopes of an element are very small, and the less abundant isotopes of an element are typically very rare, so a very sensitive instrument is required. These instruments, sometimes referred to as isotope ratio mass spectrometers (IR-MS), usually use a single magnet to bend a beam of ionized particles towards a series of Faraday cups which convert particle impacts to electric current. A fast on-line analysis of deuterium content of water can be done using flowing afterglow mass spectrometry, FA-MS. Probably the most sensitive and accurate mass spectrometer for this purpose is the accelerator mass spectrometer (AMS). This is because it provides ultimate sensitivity, capable of measuring individual atoms and measuring nuclides with a dynamic range of $\sim10^{15}$ relative to the major stable isotope. Isotope ratios are important markers of a variety of processes. Some isotope ratios are used to determine the age of materials for example as in carbon dating. Labeling with stable isotopes is also used for protein quantification.

Membrane-introduction Mass Spectrometry: Measuring Gases in Solution

Membrane-introduction mass spectrometry combines the isotope ratio MS with a reaction chamber/cell separated by a gas-permeable membrane. This method allows the study of gases as they evolve in solution. This method has been extensively used for the study of the production of oxygen by Photosystem II.

Trace Gas Analysis

Several techniques use ions created in a dedicated ion source injected into a flow tube or a drift tube: selected ion flow tube (SIFT-MS), and proton transfer reaction (PTR-MS), are variants of chemical ionization dedicated for trace gas analysis of air, breath or liquid headspace using well defined reaction time allowing calculations of analyte concentrations from the known reaction kinetics without the need for internal standard or calibration.

Atom Probe

An atom probe is an instrument that combines time-of-flight mass spectrometry and field-evaporation microscopy to map the location of individual atoms.

Pharmacokinetics

Pharmacokinetics is often studied using mass spectrometry because of the complex nature of the matrix (often blood or urine) and the need for high sensitivity to observe low dose and long time point data. The most common instrumentation used in this application is LC-MS with a triple quadrupole mass spectrometer. Tandem mass spectrometry is usually employed for added specificity. Standard curves and internal standards are used for quantitation of usually a single pharmaceutical in the samples. The samples represent different time points as a pharmaceutical is administered and then metabolized or cleared from the body. Blank or t=0 samples taken before administration are important in determining background and ensuring data integrity with such complex sample matrices. Much attention is paid to the linearity of the standard curve; however it is not uncommon to use curve fitting with more complex functions such as quadratics since the response of most mass spectrometers is less than linear across large concentration ranges.

There is currently considerable interest in the use of very high sensitivity mass spectrometry for microdosing studies, which are seen as a promising alternative to animal experimentation.

Protein Characterization

Mass spectrometry is an important method for the characterization and sequencing of proteins. The two primary methods for ionization of whole proteins are electrospray ionization (ESI) and matrix-assisted laser desorption/ionization (MALDI). In keeping with the performance and mass range of available mass spectrometers, two approaches are used for characterizing proteins. In the first, intact proteins are ionized by either of the two techniques described above, and then introduced to a mass analyzer. This approach is referred to as "top-down" strategy of protein analysis. The top-down approach however is largely limited to low-throughput single-protein studies. In the second, proteins are enzymatically digested into smaller peptides using proteases such as trypsin or pepsin, either in solution or in gel after electrophoretic separation. Other proteolytic agents are also used. The collection of peptide products are often separated by chromatography prior to introduction to the mass analyzer. When the characteristic pattern of peptides is used for the identification of the protein the method is called peptide mass fingerprinting (PMF), if the identification is performed using the sequence data determined in tandem MS analysis it is called de novo peptide sequencing. These procedures of protein analysis are also referred to as the "bottom-up" approach, and have also been used to analyse the distribution and position of post-translational modifications such as phosphorylation on proteins. A third approach is also beginning to be used, this intermediate "middle-down" approach involves analyzing proteolytic peptides that are larger than the typical tryptic peptide.

Space Exploration

As a standard method for analysis, mass spectrometers have reached other planets and moons. Two were taken to Mars by the Viking program. In early 2005 the Cassini–Huygens mission delivered a specialized GC-MS instrument aboard the Huygens probe through the atmosphere of Titan, the largest moon of the planet Saturn. This instrument analyzed atmospheric samples along its descent trajectory and was able to vaporize and analyze samples of Titan's frozen, hydrocarbon covered surface once the probe had landed. These measurements compare the abundance of isotope(s) of each particle comparatively to earth's natural abundance. Also on board the Cassini–Huygens spacecraft was an ion and neutral mass spectrometer which had been taking measurements of

Titan's atmospheric composition as well as the composition of Enceladus' plumes. A Thermal and Evolved Gas Analyzer mass spectrometer was carried by the Mars Phoenix Lander launched in 2007.

NASA's Phoenix Mars Lander analyzing a soil sample from the "Rosy Red" trench with the TEGA mass spectrometer.

Mass spectrometers are also widely used in space missions to measure the composition of plasmas. For example, the Cassini spacecraft carried the Cassini Plasma Spectrometer (CAPS), which measured the mass of ions in Saturn's magnetosphere.

Respired Gas Monitor

Mass spectrometers were used in hospitals for respiratory gas analysis beginning around 1975 through the end of the century. Some are probably still in use but none are currently being manufactured.

Found mostly in the operating room, they were a part of a complex system, in which respired gas samples from patients undergoing anesthesia were drawn into the instrument through a valve mechanism designed to sequentially connect up to 32 rooms to the mass spectrometer. A computer directed all operations of the system. The data collected from the mass spectrometer was delivered to the individual rooms for the anesthesiologist to use.

The uniqueness of this magnetic sector mass spectrometer may have been the fact that a plane of detectors, each purposely positioned to collect all of the ion species expected to be in the samples, allowed the instrument to simultaneously report all of the gases respired by the patient. Although the mass range was limited to slightly over 120 u, fragmentation of some of the heavier molecules negated the need for a higher detection limit.

Preparative Mass Spectrometry

The primary function of mass spectrometry is as a tool for chemical analyses based on detection and quantification of ions according to their mass-to-charge ratio. However, mass spectrometry also shows promise for material synthesis. Ion soft landing is characterized by deposition of intact species on surfaces at low kinetic energies which precludes the fragmentation of the incident species. The soft landing technique was first reported in 1977 for the reaction of low energy sulfur containing ions on a lead surface.

Ion-mobility Spectrometry

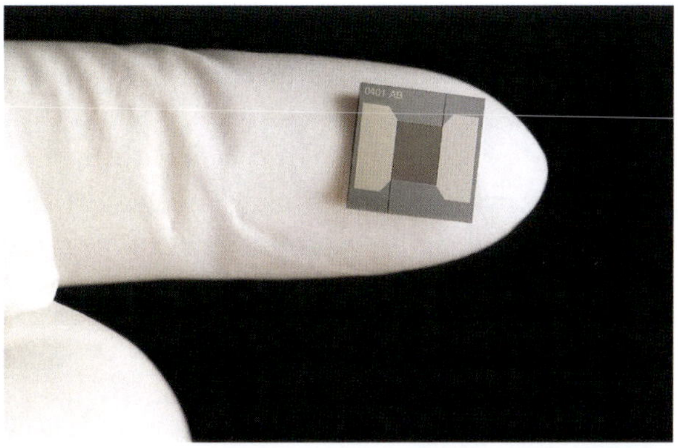

IMS chip, this dime-sized chip provides dozens of channels through which ions travel (perpendicular to plane of view) to be separated and identified.

Ion-mobility spectrometry (IMS) is an analytical technique used to separate and identify ionized molecules in the gas phase based on their mobility in a carrier buffer gas. Though heavily employed for military or security purposes, such as detecting drugs and explosives, the technique also has many laboratory analytical applications, including the analysis of both small and large biomolecules. IMS instruments are extremely sensitive stand-alone devices, but are often coupled with mass spectrometry, gas chromatography or high-performance liquid chromatography in order to achieve a multi-dimensional separation. They come in various sizes, ranging from a few millimeters to several meters depending on the specific application, and are capable of operating under a broad range of conditions. IMS instruments such as microscale high-field asymmetric waveform ion mobility spectrometry can be palm-portable for use in a range of applications including volatile organic compound (VOC) monitoring, biological sample analysis, medical diagnosis and food quality monitoring. Systems operated at higher pressure (i.e. atmospheric conditions, 1 atm or 1013 hPa) are often accompanied by elevated temperature (above 100 °C), while lower pressure systems (1-20 hPa) do not require heating.

Applications

Perhaps ion mobility spectrometry's greatest strength is the speed at which separations occur—typically on the order of tens of milliseconds. This feature combined with its ease of use, relatively high sensitivity, and highly compact design have allowed IMS as a commercial product to be used as a routine tool for the field detection of explosives, drugs, and chemical weapons. Major manufacturers of IMS screening devices used in airports are Morpho and Smiths Detection. Smiths purchased Morpho Detection in 2017 and subsequently had to legally divest ownership of the

Trace side of the business which was sold on to Rapiscan Systems in mid 2017. The products are listed under ETD Itemisers. The latest model is a non radiation 4DX.

In the pharmaceutical industry IMS is used in cleaning validations, demonstrating that reaction vessels are sufficiently clean to proceed with the next batch of pharmaceutical product. IMS is much faster and more accurate than HPLC and total organic carbon methods previously used. IMS is also used for analyzing the composition of drugs produced, thereby finding a place in quality assurance and control.

As a research tool ion mobility is becoming more widely used in the analysis of biological materials, specifically, proteomics and metabolomics. For example, IMS-MS using MALDI as the ionization method has helped make advances in proteomics, providing faster high-resolution separations of protein pieces in analysis. Moreover, it is a really promising tool for glycomics, as rotationally averaged collision cross section (CCS) values can be obtained. CCS values are important distinguishing characteristics of ions in the gas phase, and in addition to the empirical determinations it can also be calculated computationally when the 3D structure of the molecule is known. This way, adding CCS values of glycans and their fragments to databases will increase structural identification confidence and accuracy.

Outside of laboratory purposes, IMS has found great usage as a detection tool for hazardous substances. More than 10,000 IMS devices are in use worldwide in airports, and the US Army has more than 50,000 IMS devices. In industrial settings, uses of IMS include checking equipment cleanliness and detecting emission contents, such as determining the amount of hydrochloric and hydrofluoric acid in a stack gas from a process. It is also applied in industrial purposes to detect harmful substances in air.

In metabolomics the IMS is used to detect lung cancer, Chronic obstructive pulmonary disease, sarcoidosis, potential rejections after lung transplantation and relations to bacteria within the lung.

Ion Mobility

The physical quantity ion mobility K is defined as the proportionality factor between an ion's drift velocity v_d in a gas and an electric field of strength E.

$$v_d = KE$$

Ion mobilities are commonly reported as reduced mobilities, correcting to standard gas density n_o, which can be expressed in standard temperature T_o = 273 K and standard pressure p_o = 1013 hPa. This does not correct for other effects than the change in gas density and the reduced ion mobility is therefore still temperature dependent.

$$K_0 = K\frac{n}{n_0} = K\frac{T_0}{T}\frac{p}{p_0}$$

The ion mobility K can, under a variety of assumptions, be calculated by the Mason-Schamp equation,

$$K = \frac{3}{16}\sqrt{\frac{2\pi}{\mu kT}}\frac{Q}{n\sigma}$$

where Q is the ion charge, n is the drift gas number density, μ is the reduced mass of the ion and the drift gas molecules, k is Boltzmann constant, T is the drift gas temperature, and σ is the collision cross section between the ion and the drift gas molecules. Often, N is used instead of n for the drift gas number density and Ω instead σ for the ion-neutral collision cross section. This relation holds approximately at a low electric field limit, where the ratio of E/N is small and thus the thermal energy of the ions is much greater than the energy gained from the electric field between collisions. With these ions having similar energies as the buffer gas molecules, diffusion forces dominate ion motion in this case. The ratio E/N is typically given in Townsends (Td) and the transition between low- and high-field conditions is typically estimated to occur between 2 Td and 10 Td. When low-field conditions no longer prevail, the ion mobility itself becomes a function of the electric field strength which is usually described empirically through the so-called alpha function.

$$K(\frac{E}{N}) = K(0)[1+\alpha(\frac{E}{N})] = K(0)[1+\alpha_2[\frac{E}{N}]^2 + \alpha_4[\frac{E}{N}]^4 + ...]$$

Ionization

The molecules of the sample need to be ionized, usually by corona discharge, atmospheric pressure photoionization (APPI), electrospray ionization (ESI), or radioactive atmospheric-pressure chemical ionization (R-APCI) source, e.g. a small piece of ^{63}Ni or ^{241}Am, similar to the one used in ionization smoke detectors. ESI and MALDI techniques are commonly used when IMS is paired with mass spectrometry.

Doping materials are sometimes added to the drift gas for ionization selectivity. For example, acetone can be added for chemical warfare agent detection, chlorinated solvents added for explosives, and nicotinamide added for drugs detection.

Analyzers

Ion mobility spectrometers exist based on various principles, optimized for different applications.

Drift Tube

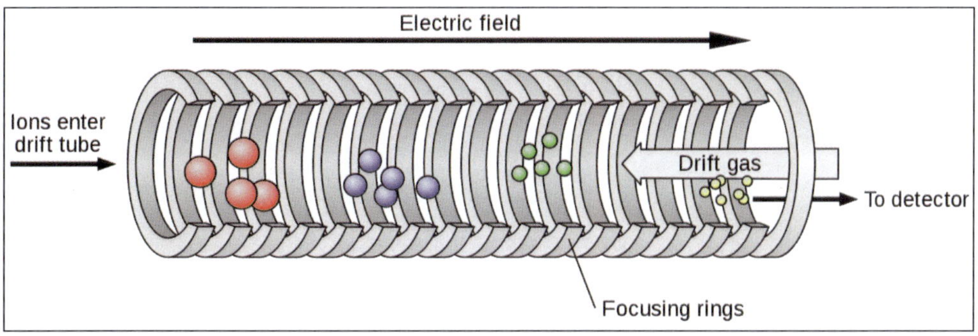

A drift tube ion mobility spectrometer.

In its simplest form, an IMS system measures how long a given ion takes to traverse a given length in a uniform electric field through a given atmosphere. In specified intervals, a sample of the ions is let into the drift chamber; the gating mechanism is based on a charged electrode working in a similar way as the control grid in triodes works for electrons. For precise control of the ion pulse width admitted to the drift tube, more complex gating systems such as a Bradbury-Nielsen or a Field Switching Shutter are employed. Once in the drift tube, ions are subjected to a homogeneous electric field ranging from a few volts per centimeter up to many hundreds of volts per centimeter. This electric field then drives the ions through the drift tube where they interact with the neutral drift molecules contained within the system and separate based on the ion mobility, arriving at the detector for measurement. Ions are recorded at the detector in order from the fastest to the slowest, generating a response signal characteristic for the chemical composition of the measured sample.

The ion mobility K can then be experimentally determined from the drift time t_D of an ion traversing within a homogeneous electric field the potential difference U in the drift length L,

$$K = \frac{L^2}{t_D U}$$

A drift tube's resolving power R_p can, when diffusion is assumed as the sole contributor to peak broadening, be calculated as,

$$R_P = \frac{t_D}{\Delta t_D} = \sqrt{\frac{LEQ}{16kT \ln 2}}$$

where t_D is the ion drift time, Δt_D is the Full width at half maximum, L is the tube length, E is the electric field strength, Q is the ion charge, k is Boltzmann's constant, and T is the drift gas temperature. Ambient pressure methods allow for higher resolving power and greater separation selectivity due to a higher rate of ion-molecule interactions and is typically used for stand-alone devices, as well as for detectors for gas, liquid, and supercriticial fluid chromatography. The resolving power depends on the total voltage drop the ion traverses. Using a drift voltage of 25 kV in a 15 cm long atmospheric pressure drift tube, a resolving power above 250 is achievable even for small, single charged ions. This is sufficient to achieve separation of some isotopologues based on their difference in reduced mass μ.

Low Pressure Drift Tube

Reduced pressure drift tubes operate using the same principles as their atmospheric pressure counterparts, but at drift gas pressure of only a few torr. Due to the vastly reduced number of ion-neutral interactions, much longer drift tubes or much faster ion shutters are necessary to achieve the same resolving power. However, the reduced pressure operation offers several advantages. First, it eases interfacing the IMS with mass spectrometry. Second, at lower pressures, ions can be stored for injection from an ion trap and re-focussed radially during and after the separation. Third, high values of E/N can be achieved, allowing for direct measurement of $K(E/N)$ over a wide range.

Travelling Wave

Though drift electric fields are normally uniform, non-uniform drift fields can also be used. One example is the travelling wave IMS, which is a low pressure drift tube IMS where the electric field is only applied in a small region of the drift tube. This region then moves along the drift tube, creating a wave pushing the ions towards the detector, removing the need for a high total drift voltage.

An especially noteworthy variant is the "SUPER" IMS, which combines ion trapping by the so-called structures for lossless ion manipulations (SLIM) with several passes through the same drift region to achieve extremely high resolving powers.

Trapped Ion Mobility Spectrometry

In trapped ion mobility spectrometry (TIMS), ions are held stationary (or trapped) in a flowing buffer gas by an axial electric field gradient (EFG) profile while the application of radio frequency (rf) potentials results in trapping in the radial dimension. TIMS operates in the pressure range of 2 to 5 hPa and replaces the ion funnel found in the source region of modern mass spectrometers. It can be coupled with nearly any mass analyzer through either the standard mode of operation for beam-type instruments or selective accumulation mode (SA-TIMS) when used with trapping mass spectrometry (MS) instruments.

Effectively, the drift cell is prolonged by the ion motion created through the gas flow. Thus, TIMS devices do neither require large size nor high voltage in order to achieve high resolution, for instance achieving over 250 resolving power from a 4.7 cm device through the use of extended separation times. However, the resolving power strongly depends on the ion mobility and decreases for more mobile ions. In addition, TIMS can be capable of higher sensitivity than other ion mobility systems because no grids or shutters exist in the ion path, improving ion transmission both during ion mobility experiments and while operating in a transparent MS only mode.

High-field Asymmetric Waveform Ion Mobility Spectrometry

DMS (differential mobility spectrometer) or FAIMS (field asymmetric ion mobility spectrometer) make use of the dependence of the ion mobility K on the electric field strength E at high electric fields. Ions are transported through the device by the drift gas flow and subjected to different field strengths in orthogonal direction for different amounts of time. Ions are deflected towards the walls of the analyzer based on the change of their mobility. Thereby only ions with a certain mobility dependence can pass the thus created filter.

Differential Mobility Analyzer

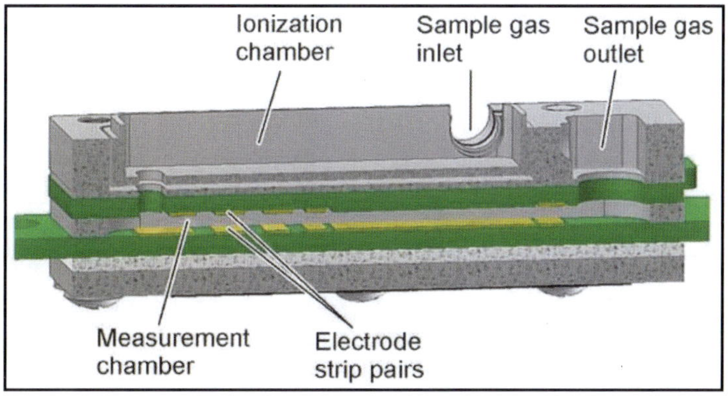

Example of Aspiration IMS sensor.

A differential mobility analyzer (DMA) makes use of a fast gas stream perpendicular to the electric field. Thereby ions of different mobilities undergo different trajectories. This type of IMS corresponds to the sector instruments in mass spectrometry. They also work as a scannable filter. Examples include the differential mobility detector first commercialized by Varian in the CP-4900 MicroGC. Aspiration IMS operates with open-loop circulation of sampled air. Sample flow is passed via ionization chamber and then enters to measurement area where the ions are deflected into one or more measuring electrodes by perpendicular electric field which can be either static or varying. The output of the sensor is characteristic of the ion mobility distribution and can be used for detection and identification purposes.

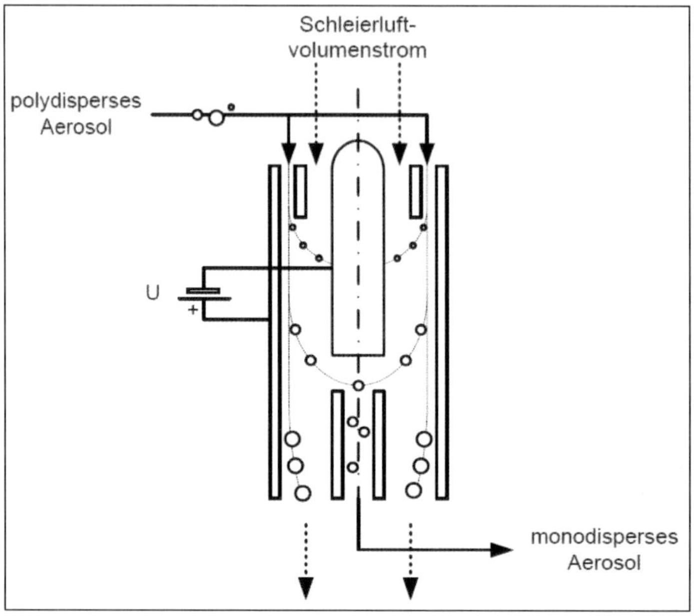

Principle of operation of a differential mobility analyzer for aerosol separation.

A DMA can separate charged aerosol particles or ions according to their mobility in an electric field prior to their detection, which can be done with several means, including electrometers or the more sophisticated mass spectrometers.

Drift Gas

The drift gas composition is an important parameter for the IMS instrument design and resolution. Often, different drift gas compositions can allow for the separation of otherwise overlapping peaks. Elevated gas temperature assists in removing ion clusters that may distort experimental measurements.

Detector

Often the detector is a simple Faraday plate coupled to a transimpedance amplifier, however, more advanced ion mobility instruments are coupled with mass spectrometers in order to obtain both size and mass information simultaneously. It is noteworthy that the detector influences the optimum operating conditions for the ion mobility experiment.

Combined Methods

IMS can be combined with other separation techniques.

Gas Chromatography

When IMS is coupled with gas chromatography, common sample introduction is with the GC capillary column directly connected to the IMS setup, with molecules ionized as they elute from GC. A similar technique is commonly used for HPLC. A novel design for corona discharge ionization ion mobility spectrometry (CD–IMS) as a detector after capillary gas chromatography has been produced in 2012. In this design, a hollow needle was used for corona discharge creation and the effluent was entered into the ionization region on the upstream side of the corona source. In addition to the practical conveniences in coupling the capillary to IMS cell, this direct axial interfacing helps us to achieve a more efficient ionization, resulting in higher sensitivity.

When used with GC, a differential mobility analyzer is often called a differential mobility detector (DMD). A DMD is often a type of microelectromechanical system, radio frequency modulated ion mobility spectrometry (MEMS RF-IMS) device. Though small, it can fit into portable units, such as transferable gas chromatographs or drug/explosives sensors. For instance, it was incorporated by Varian in its CP-4900 DMD MicroGC, and by Thermo Fisher in its EGIS Defender system, designed to detect narcotics and explosives in transportation or other security applications.

Liquid Chromatography

Coupled with LC and MS, IMS has become widely used to analyze biomolecules, a practice heavily developed by David E. Clemmer, now at Indiana University (Bloomington).

Mass Spectrometry

When IMS is used with mass spectrometry, ion mobility spectrometry-mass spectrometry offers many advantages, including better signal to noise, isomer separation, and charge state identification. IMS has commonly been attached to several mass spec analyzers, including quadropole, time-of-flight, and Fourier transform cyclotron resonance.

Fluorescence Spectroscopy

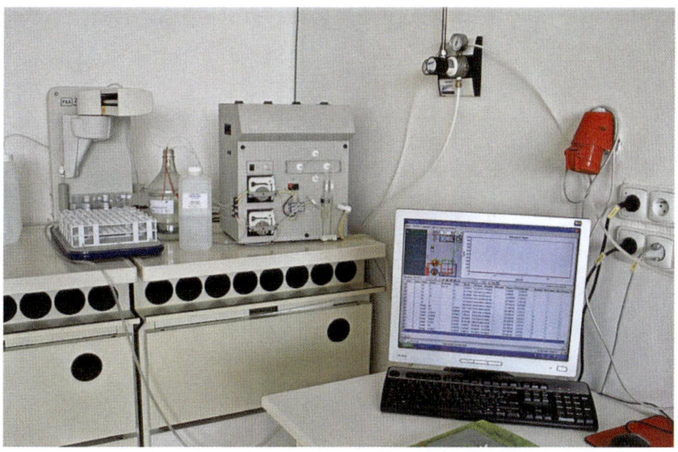

Atomic fluorescence spectroscopy analyzer for of mercury.

Fluorescence spectroscopy (also known as fluorimetry or spectrofluorometry) is a type of electromagnetic spectroscopy that analyzes fluorescence from a sample. It involves using a beam of light, usually ultraviolet light, that excites the electrons in molecules of certain compounds and causes them to emit light; typically, but not necessarily, visible light. A complementary technique is absorption spectroscopy. In the special case of single molecule fluorescence spectroscopy, intensity fluctuations from the emitted light are measured from either single fluorophores, or pairs of fluorophores. Devices that measure fluorescence are called fluorometers.

Molecules have various states referred to as energy levels. Fluorescence spectroscopy is primarily concerned with electronic and vibrational states. Generally, the species being examined has a ground electronic state (a low energy state) of interest, and an excited electronic state of higher energy. Within each of these electronic states there are various vibrational states.

In fluorescence, the species is first excited, by absorbing a photon, from its ground electronic state to one of the various vibrational states in the excited electronic state. Collisions with other molecules cause the excited molecule to lose vibrational energy until it reaches the lowest vibrational state from the excited electronic state.

The molecule then drops down to one of the various vibrational levels of the ground electronic state again, emitting a photon in the process. As molecules may drop down into any of several vibrational levels in the ground state, the emitted photons will have different energies, and thus frequencies. Therefore, by analysing the different frequencies of light emitted in fluorescent spectroscopy, along with their relative intensities, the structure of the different vibrational levels can be determined.

For atomic species, the process is similar; however, since atomic species do not have vibrational energy levels, the emitted photons are often at the same wavelength as the incident radiation. This process of re-emitting the absorbed photon is "resonance fluorescence" and while it is characteristic of atomic fluorescence, is seen in molecular fluorescence as well.

In a typical fluorescence (emission) measurement, the excitation wavelength is fixed and the detection wavelength varies, while in a fluorescence excitation measurement the detection wavelength is fixed and the excitation wavelength is varied across a region of interest. An emission map is measured by recording the emission spectra resulting from a range of excitation wavelengths and combining them all together. This is a three dimensional surface data set: emission intensity as a function of excitation and emission wavelengths, and is typically depicted as a contour map.

Instrumentation

Two general types of instruments exist: filter fluorometers that use filters to isolate the incident light and fluorescent light and spectrofluorometers that use a diffraction grating monochromators to isolate the incident light and fluorescent light.

Both types use the following scheme: the light from an excitation source passes through a filter or monochromator, and strikes the sample. A proportion of the incident light is absorbed by the sample, and some of the molecules in the sample fluoresce. The fluorescent light is emitted in all directions. Some of this fluorescent light passes through a second filter or monochromator and reaches a detector, which is usually placed at 90° to the incident light beam to minimize the risk of transmitted or reflected incident light reaching the detector.

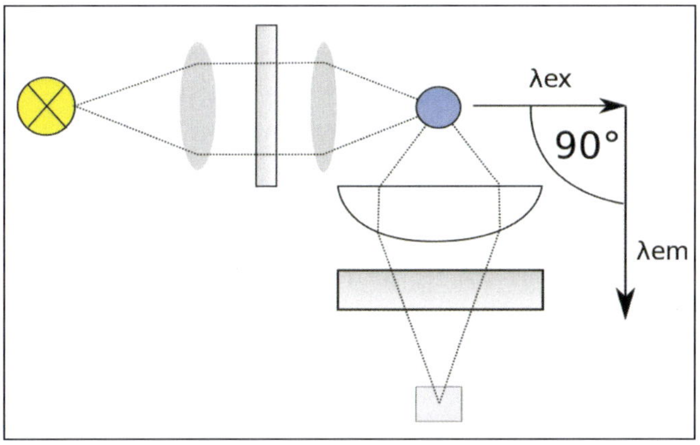

A simplistic design of the components of a fluorimeter.

Various light sources may be used as excitation sources, including lasers, LED, and lamps; xenon arcs and mercury-vapor lamps in particular. A laser only emits light of high irradiance at a very narrow wavelength interval, typically under 0.01 nm, which makes an excitation monochromator or filter unnecessary. The disadvantage of this method is that the wavelength of a laser cannot be changed by much. A mercury vapor lamp is a line lamp, meaning it emits light near peak wavelengths. By contrast, a xenon arc has a continuous emission spectrum with nearly constant intensity in the range from 300-800 nm and a sufficient irradiance for measurements down to just above 200 nm.

Filters and monochromators may be used in fluorimeters. A monochromator transmits light of an adjustable wavelength with an adjustable tolerance. The most common type of monochromator utilizes a diffraction grating, that is, collimated light illuminates a grating and exits with a different angle depending on the wavelength. The monochromator can then be adjusted to select which wavelengths to transmit. For allowing anisotropy measurements, the addition of two polarization filters is necessary: one after the excitation monochromator or filter, and one before the emission monochromator or filter.

The fluorescence is most often measured at a 90° angle relative to the excitation light. This geometry is used instead of placing the sensor at the line of the excitation light at a 180° angle in order to avoid interference of the transmitted excitation light. No monochromator is perfect and it will transmit some stray light, that is, light with other wavelengths than the targeted. An ideal monochromator would only transmit light in the specified range and have a high wavelength-independent transmission. When measuring at a 90° angle, only the light scattered by the sample causes stray light. This results in a better signal-to-noise ratio, and lowers the detection limit by approximately a factor 10000, when compared to the 180° geometry. Furthermore, the fluorescence can also be measured from the front, which is often done for turbid or opaque samples.

The detector can either be single-channeled or multichanneled. The single-channeled detector can only detect the intensity of one wavelength at a time, while the multichanneled detects the intensity of all wavelengths simultaneously, making the emission monochromator or filter unnecessary. The different types of detectors have both advantages and disadvantages.

The most versatile fluorimeters with dual monochromators and a continuous excitation light source can record both an excitation spectrum and a fluorescence spectrum. When measuring fluorescence spectra, the wavelength of the excitation light is kept constant, preferably at a wavelength of high absorption, and the emission monochromator scans the spectrum. For measuring excitation spectra, the wavelength passing through the emission filter or monochromator is kept constant and the excitation monochromator is scanning. The excitation spectrum generally is identical to the absorption spectrum as the fluorescence intensity is proportional to the absorption.

Analysis of Data

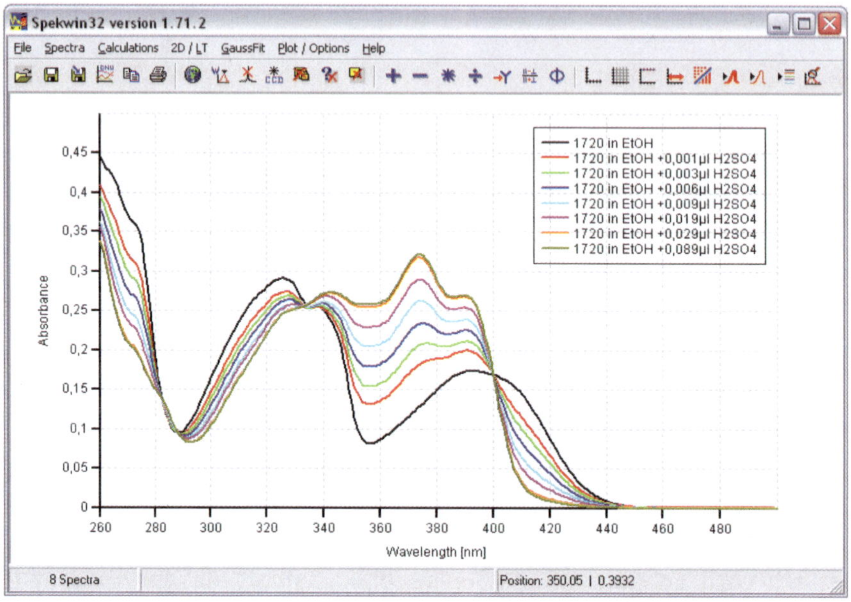

At low concentrations the fluorescence intensity will generally be proportional to the concentration of the fluorophore.

Unlike in UV/visible spectroscopy, 'standard', device independent spectra are not easily attained. Several factors influence and distort the spectra, and corrections are necessary to attain 'true', i.e. machine-independent, spectra. The different types of distortions will here be classified as being either instrument- or sample-related. Firstly, the distortion arising from the instrument is discussed. As a start, the light source intensity and wavelength characteristics varies over time during each experiment and between each experiment. Furthermore, no lamp has a constant intensity at all wavelengths. To correct this, a beam splitter can be applied after the excitation monochromator or filter to direct a portion of the light to a reference detector.

Additionally, the transmission efficiency of monochromators and filters must be taken into account. These may also change over time. The transmission efficiency of the monochromator also varies depending on wavelength. This is the reason that an optional reference detector should be placed after the excitation monochromator or filter. The percentage of the fluorescence picked up by the detector is also dependent upon the system. Furthermore, the detector quantum efficiency, that is, the percentage of photons detected, varies between different detectors, with wavelength and with time, as the detector inevitably deteriorates.

Two other topics that must be considered include the optics used to direct the radiation and the means of holding or containing the sample material (called a cuvette or cell). For most UV, visible, and NIR measurements the use of precision quartz cuvettes is necessary. In both cases, it is important to select materials that have relatively little absorption in the wavelength range of interest. Quartz is ideal because it transmits from 200 nm-2500 nm; higher grade quartz can even transmit up to 3500 nm, whereas the absorption properties of other materials can mask the fluorescence from the sample.

Correction of all these instrumental factors for getting a 'standard' spectrum is a tedious process, which is only applied in practice when it is strictly necessary. This is the case when measuring the quantum yield or when finding the wavelength with the highest emission intensity for instance.

As mentioned earlier, distortions arise from the sample as well. Therefore, some aspects of the sample must be taken into account too. Firstly, photodecomposition may decrease the intensity of fluorescence over time. Scattering of light must also be taken into account. The most significant types of scattering in this context are Rayleigh and Raman scattering. Light scattered by Rayleigh scattering has the same wavelength as the incident light, whereas in Raman scattering the scattered light changes wavelength usually to longer wavelengths. Raman scattering is the result of a virtual electronic state induced by the excitation light. From this virtual state, the molecules may relax back to a vibrational level other than the vibrational ground state. In fluorescence spectra, it is always seen at a constant wavenumber difference relative to the excitation wavenumber e.g. the peak appears at a wavenumber 3600 cm^{-1} lower than the excitation light in water.

Other aspects to consider are the inner filter effects. These include reabsorption. Reabsorption happens because another molecule or part of a macromolecule absorbs at the wavelengths at which the fluorophore emits radiation. If this is the case, some or all of the photons emitted by the fluorophore may be absorbed again. Another inner filter effect occurs because of high concentrations of absorbing molecules, including the fluorophore. The result is that the intensity of the excitation light is not constant throughout the solution. Resultingly, only a small percentage of the excitation light reaches the fluorophores that are visible for the detection system. The inner filter effects change the spectrum and intensity of the emitted light and they must therefore be considered when analysing the emission spectrum of fluorescent light.

Tryptophan Fluorescence

The fluorescence of a folded protein is a mixture of the fluorescence from individual aromatic residues. Most of the intrinsic fluorescence emissions of a folded protein are due to excitation of tryptophan residues, with some emissions due to tyrosine and phenylalanine; but disulfide bonds also have appreciable absorption in this wavelength range. Typically, tryptophan has a wavelength of maximum absorption of 280 nm and an emission peak that is solvatochromic, ranging from ca. 300 to 350 nm depending in the polarity of the local environment. Hence, protein fluorescence may be used as a diagnostic of the conformational state of a protein. Furthermore, tryptophan fluorescence is strongly influenced by the proximity of other residues (*i.e.*, nearby protonated groups such as Asp or Glu can cause quenching of Trp fluorescence). Also, energy transfer between tryptophan and the other fluorescent amino acids is possible, which would affect the analysis, especially in cases where the Förster acidic approach is taken. In addition, tryptophan is a relatively rare amino acid; many proteins contain only one or a few tryptophan residues. Therefore, tryptophan fluorescence can be a very sensitive measurement of the conformational state of individual tryptophan residues. The advantage compared to extrinsic probes is that the protein itself is not changed. The use of intrinsic fluorescence for the study of protein conformation is in practice limited to cases with few (or perhaps only one) tryptophan residues, since each experiences a different local environment, which gives rise to different emission spectra.

Tryptophan is an important intrinsic fluorescent (amino acid), which can be used to estimate the nature of microenvironment of the tryptophan. When performing experiments with denaturants, surfactants or other amphiphilic molecules, the microenvironment of the tryptophan might change. For example, if a protein containing a single tryptophan in its 'hydrophobic' core is denatured with increasing temperature, a red-shifted emission spectrum will appear. This is due to the exposure of the tryptophan to an aqueous environment as opposed to a hydrophobic protein interior. In contrast, the addition of a surfactant to a protein which contains a tryptophan which is exposed to the aqueous solvent will cause a blue-shifted emission spectrum if the tryptophan is embedded in the surfactant vesicle or micelle. Proteins that lack tryptophan may be coupled to a fluorophore.

With fluorescence excitation at 295 nm, the tryptophan emission spectrum is dominant over the weaker tyrosine and phenylalanine fluorescence.

Applications

Fluorescence spectroscopy is used in, among others, biochemical, medical, and chemical research fields for analyzing organic compounds. There has also been a report of its use in differentiating malignant skin tumors from benign. Atomic Fluorescence Spectroscopy (AFS) techniques are useful in other kinds of analysis/measurement of a compound present in air or water, or other media, such as CVAFS which is used for heavy metals detection, such as mercury. Fluorescence can also be used to redirect photons. Additionally, Fluorescence spectroscopy can be adapted to the microscopic level using microfluorimetry. In analytical chemistry, fluorescence detectors are used with HPLC.

Electroanalytical Methods

Electroanalytical methods are a class of techniques in analytical chemistry which study an analyte by measuring the potential (volts) and current (amperes) in an electrochemical cell containing the analyte. These methods can be broken down into several categories depending on which aspects of the cell are controlled and which are measured. The three main categories are potentiometry (the difference in electrode potentials is measured), coulometry (the cell's current is measured over time), and voltammetry (the cell's current is measured while actively altering the cell's potential).

Potentiometry

Potentiometry passively measures the potential of a solution between two electrodes, affecting the solution very little in the process. One electrode is called the reference electrode and has a constant potential, while the other one is an indicator electrode whose potential changes with the composition of the sample. Therefore, the difference of potential between the two electrodes gives an assessment of the composition of the sample. In fact, since potentiometric measurement is a non-destructive measurement, assuming that the electrode is in equilibrium with the solution we are measuring the potential of the solution. Potentiometry usually uses indicator electrodes made selectively sensitive to the ion of interest, such as fluoride in fluoride selective electrodes, so that the potential solely depends on the activity of this ion of interest. The time that takes the electrode to establish equilibrium with the solution will affect the sensitivity or accuracy of the measurement. In aquatic environments, platinum is often used due to its high electron transfer kinetics, although an electrode made from several metals can be used in order to enhance to electron transfer kinetics. The most common potentiometric electrode is by far the glass-membrane electrode used in a pH meter.

A variant of potentiometry is chronopotentiometry which consists in using a constant current and measurement of potential as a function of time. It has been initiated by Weber.

Coulometry

Coulometry uses applied current or potential to completely convert an analyte from one oxidation state to another. In these experiments, the total current passed is measured directly or indirectly to determine the number of electrons passed. Knowing the number of electrons passed can indicate the concentration of the analyte or, when the concentration is known, the number of electrons transferred in the redox reaction. Common forms of coulometry include bulk electrolysis, also known as Potentiostatic coulometry or controlled potential coulometry, as well as a variety of coulometric titrations.

Voltammetry

Voltammetry applies a constant and varying potential at an electrode's surface and measures the resulting current with a three electrode system. This method can reveal the reduction potential of an analyte and its electrochemical reactivity. This method in practical terms is nondestructive since only a very small amount of the analyte is consumed at the two-dimensional surface of the working and auxiliary electrodes. In practice the analyte solutions is usually disposed of since it is difficult to separate the analyte from the bulk electrolyte and the experiment requires a small amount of analyte. A normal experiment may involve 1–10 mL solution with an analyte concentration between 1 and 10 mmol/L. Chemically modified electrodes are employed for analysis of organic and inorganic samples.

Polarography

Polarography is a subclass of voltammetry that uses a dropping mercury electrode as the working electrode.

Amperometry

Amperometry is the term indicating the whole of electrochemical techniques in which a current is measured as a function of an independent variable that is, typically, time or electrode potential. Chronoamperometry is the technique in which the current is measured, at a fixed potential, at different times since the start of polarisation. Chronoamperometry is typically carried out in unstirred solution and at fixed electrode, i.e., under experimental conditions avoiding convection as the mass transfer to the electrode. On the other hand, voltammetry is a subclass of amperometry, in which the current is measured by varying the potential applied to the electrode. According to the waveform that describes the way how the potential is varied as a function of time, the different voltammetric techniques are defined. Confusion arose recently about the correct use of many terms proper of electrochemistry/electroanalysis, often owing to the diffusion of electroanalytical techniques in fields where they constitute an instrument to use, not being the 'core business' of the study. Though electrochemists are pleased about this, they invite to use the terms properly, in order to avoid fatal misunderstandings.

Chromatography

Thin-layer chromatography is used to separate components of a plant extract, illustrating the experiment with plant pigments which gave chromatography its name.

Chromatography is a laboratory technique for the separation of a mixture. The mixture is dissolved in a fluid called the *mobile phase,* which carries it through a structure holding another material called the *stationary phase.* The various constituents of the mixture travel at different speeds, causing them to separate. The separation is based on differential partitioning between the mobile and stationary phases. Subtle differences in a compound's partition coefficient result in differential retention on the stationary phase and thus affect the separation.

Chromatography may be preparative or analytical. The purpose of preparative chromatography is to separate the components of a mixture for later use, and is thus a form of purification. Analytical chromatography is done normally with smaller amounts of material and is for establishing the presence or measuring the relative proportions of analytes in a mixture. The two are not mutually exclusive.

Chromatography Terms

- The analyte is the substance to be separated during chromatography. It is also normally what is needed from the mixture.
- Analytical chromatography is used to determine the existence and possibly also the concentration of analyte(s) in a sample.
- A bonded phase is a stationary phase that is covalently bonded to the support particles or to the inside wall of the column tubing.
- A *chromatogram* is the visual output of the chromatograph. In the case of an optimal separation, different peaks or patterns on the chromatogram correspond to different components of the separated mixture.

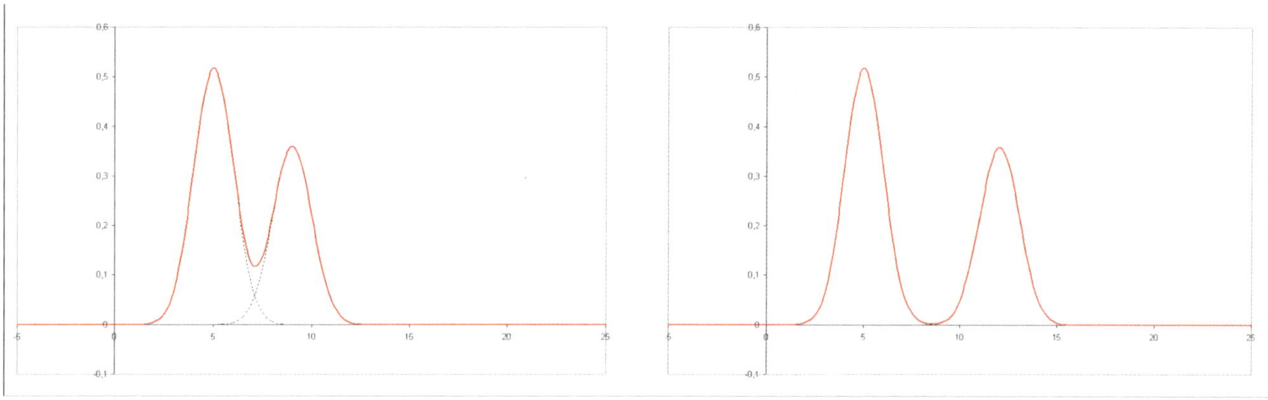

Plotted on the x-axis is the retention time and plotted on the y-axis a signal (for example obtained by a spectrophotometer, mass spectrometer or a variety of other detectors) corresponding to the response created by the analytes exiting the system. In the case of an optimal system the signal is proportional to the concentration of the specific analyte separated.

- A chromatograph is equipment that enables a sophisticated separation, e.g. gas chromatographic or liquid chromatographic separation.
- Chromatography is a physical method of separation that distributes components to separate between two phases, one stationary (stationary phase), the other (the mobile phase) moving in a definite direction.
- The eluate is the mobile phase leaving the column. This is also called effluent.
- The eluent is the solvent that carries the analyte.
- The eluite is the analyte, the eluted solute.
- An eluotropic series is a list of solvents ranked according to their eluting power.
- An immobilized phase is a stationary phase that is immobilized on the support particles, or on the inner wall of the column tubing.

- The mobile phase is the phase that moves in a definite direction. It may be a liquid (LC and Capillary Electrochromatography (CEC)), a gas (GC), or a supercritical fluid (supercritical-fluid chromatography, SFC). The mobile phase consists of the sample being separated/analyzed and the solvent that moves the sample through the column. In the case of HPLC the mobile phase consists of a non-polar solvent(s) such as hexane in normal phase or a polar solvent such as methanol in reverse phase chromatography and the sample being separated. The mobile phase moves through the chromatography column (the stationary phase) where the sample interacts with the stationary phase and is separated.

- Preparative chromatography is used to purify sufficient quantities of a substance for further use, rather than analysis.

- The retention time is the characteristic time it takes for a particular analyte to pass through the system (from the column inlet to the detector) under set conditions.

- The sample is the matter analyzed in chromatography. It may consist of a single component or it may be a mixture of components. When the sample is treated in the course of an analysis, the phase or the phases containing the analytes of interest is referred to as the sample whereas everything out of interest separated from the sample before or in the course of the analysis is referred to as waste.

- The solute refers to the sample components in partition chromatography.

- The solvent refers to any substance capable of solubilizing another substance, and especially the liquid mobile phase in liquid chromatography.

- The stationary phase is the substance fixed in place for the chromatography procedure. Examples include the silica layer in thin-layer chromatography.

- The detector refers to the instrument used for qualitative and quantitative detection of analytes after separation.

Chromatography is based on the concept of partition coefficient. Any solute partitions between two immiscible solvents. When we make one solvent immobile (by adsorption on a solid support matrix) and another mobile it results in most common applications of chromatography. If the matrix support, or stationary phase, is polar (e.g. paper, silica etc.) it is forward phase chromatography, and if it is non-polar (C-18) it is reverse phase.

Techniques by Chromatographic Bed Shape

Column Chromatography

Column chromatography is a separation technique in which the stationary bed is within a tube. The particles of the solid stationary phase or the support coated with a liquid stationary phase may fill the whole inside volume of the tube (packed column) or be concentrated on or along the inside tube wall leaving an open, unrestricted path for the mobile phase in the middle part of the tube (open tubular column). Differences in rates of movement through the medium are calculated to different retention times of the sample.

In 1978, W. Clark Still introduced a modified version of column chromatography called *flash column chromatography* (flash). The technique is very similar to the traditional column chromatography, except for that the solvent is driven through the column by applying positive pressure. This allowed most separations to be performed in less than 20 minutes, with improved separations compared to the old method. Modern flash chromatography systems are sold as pre-packed plastic cartridges, and the solvent is pumped through the cartridge. Systems may also be linked with detectors and fraction collectors providing automation. The introduction of gradient pumps resulted in quicker separations and less solvent usage.

In expanded bed adsorption, a fluidized bed is used, rather than a solid phase made by a packed bed. This allows omission of initial clearing steps such as centrifugation and filtration, for culture broths or slurries of broken cells.

Phosphocellulose chromatography utilizes the binding affinity of many DNA-binding proteins for phosphocellulose. The stronger a protein's interaction with DNA, the higher the salt concentration needed to elute that protein.

Planar Chromatography

Planar chromatography is a separation technique in which the stationary phase is present as or on a plane. The plane can be a paper, serving as such or impregnated by a substance as the stationary bed (paper chromatography) or a layer of solid particles spread on a support such as a glass plate (thin-layer chromatography). Different compounds in the sample mixture travel different distances according to how strongly they interact with the stationary phase as compared to the mobile phase. The specific Retention factor (R_f) of each chemical can be used to aid in the identification of an unknown substance.

Paper Chromatography

Paper chromatography in progress.

Paper chromatography is a technique that involves placing a small dot or line of sample solution onto a strip of *chromatography paper*. The paper is placed in a container with a shallow layer of solvent and sealed. As the solvent rises through the paper, it meets the sample mixture, which starts to travel up the paper with the solvent. This paper is made of cellulose, a polar substance, and the compounds within the mixture travel farther if they are non-polar. More polar substances bond with the cellulose paper more quickly, and therefore do not travel as far.

Thin-layer Chromatography

Thin-layer chromatography (TLC) is a widely employed laboratory technique used to separate different biochemicals on the basis of their relative attractions to the stationary and mobile phases. It is similar to paper chromatography. However, instead of using a stationary phase of paper, it involves a stationary phase of a thin layer of adsorbent like silica gel, alumina, or cellulose on a flat, inert substrate. TLC is very versatile; multiple samples can be separated simultaneously on the same layer, making it very useful for screening applications such as testing drug levels and water purity. Possibility of cross-contamination is low since each separation is performed on a new layer. Compared to paper, it has the advantage of faster runs, better separations, better quantitative analysis, and the choice between different adsorbents. For even better resolution and faster separation that utilizes less solvent, high-performance TLC can be used. An older popular use had been to differentiate chromosomes by observing distance in gel (separation of was a separate step).

Displacement Chromatography

The basic principle of displacement chromatography is: A molecule with a high affinity for the chromatography matrix (the displacer) competes effectively for binding sites, and thus displaces all molecules with lesser affinities. There are distinct differences between displacement and elution chromatography. In elution mode, substances typically emerge from a column in narrow, Gaussian peaks. Wide separation of peaks, preferably to baseline, is desired for maximum purification. The speed at which any component of a mixture travels down the column in elution mode depends on many factors. But for two substances to travel at different speeds, and thereby be resolved, there must be substantial differences in some interaction between the biomolecules and the chromatography matrix. Operating parameters are adjusted to maximize the effect of this difference. In many cases, baseline separation of the peaks can be achieved only with gradient elution and low column loadings. Thus, two drawbacks to elution mode chromatography, especially at the preparative scale, are operational complexity, due to gradient solvent pumping, and low throughput, due to low column loadings. Displacement chromatography has advantages over elution chromatography in that components are resolved into consecutive zones of pure substances rather than "peaks". Because the process takes advantage of the nonlinearity of the isotherms, a larger column feed can be separated on a given column with the purified components recovered at significantly higher concentrations.

Techniques by Physical State of Mobile Phase

Gas Chromatography

Gas chromatography (GC), also sometimes known as gas-liquid chromatography, (GLC), is a separation technique in which the mobile phase is a gas. Gas chromatographic separation is always carried out in a column, which is typically "packed" or "capillary". Packed columns are the routine work horses of gas chromatography, being cheaper and easier to use and often giving adequate performance. Capillary columns generally give far superior resolution and although more expensive are becoming widely used, especially for complex mixtures. Both types of column are made from non-adsorbent and chemically inert materials. Stainless steel and glass are the usual materials for packed columns and quartz or fused silica for capillary columns.

Gas chromatography is based on a partition equilibrium of analyte between a solid or viscous liquid stationary phase (often a liquid silicone-based material) and a mobile gas (most often helium). The stationary phase is adhered to the inside of a small-diameter (commonly 0.53 – 0.18 mm inside diameter) glass or fused-silica tube (a capillary column) or a solid matrix inside a larger metal tube (a packed column). It is widely used in analytical chemistry; though the high temperatures used in GC make it unsuitable for high molecular weight biopolymers or proteins (heat denatures them), frequently encountered in biochemistry, it is well suited for use in the petrochemical, environmental monitoring and remediation, and industrial chemical fields. It is also used extensively in chemistry research.

Liquid Chromatography

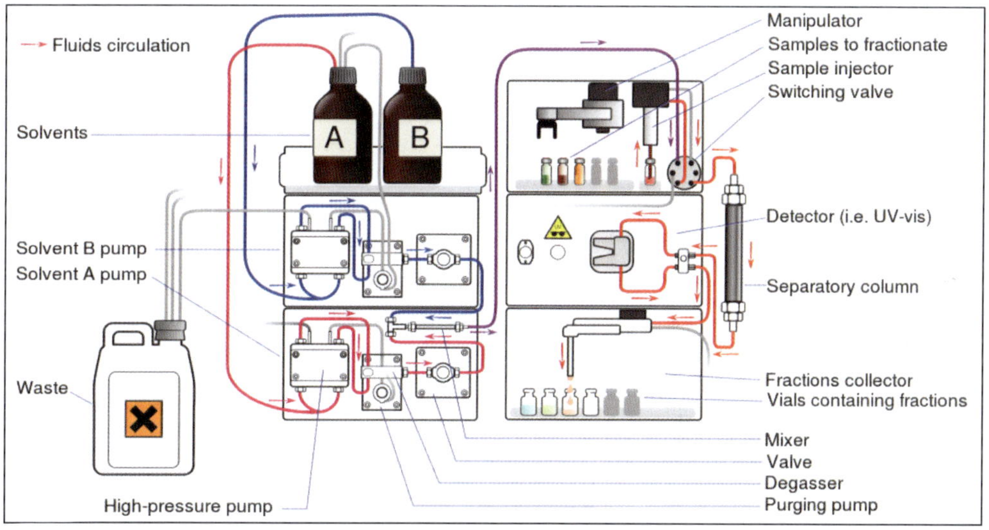

Preparative HPLC apparatus.

Liquid chromatography (LC) is a separation technique in which the mobile phase is a liquid. It can be carried out either in a column or a plane. Present day liquid chromatography that generally utilizes very small packing particles and a relatively high pressure is referred to as high-performance liquid chromatography (HPLC).

In HPLC the sample is forced by a liquid at high pressure (the mobile phase) through a column that is packed with a stationary phase composed of irregularly or spherically shaped particles, a porous monolithic layer, or a porous membrane. HPLC is historically divided into two different sub-classes based on the polarity of the mobile and stationary phases. Methods in which the stationary phase is more polar than the mobile phase (e.g., toluene as the mobile phase, silica as the stationary phase) are termed normal phase liquid chromatography (NPLC) and the opposite (e.g., water-methanol mixture as the mobile phase and C18 (octadecylsilyl) as the stationary phase) is termed reversed phase liquid chromatography (RPLC).

Affinity Chromatography

Affinity chromatography is based on selective non-covalent interaction between an analyte and specific molecules. It is very specific, but not very robust. It is often used in biochemistry in the purification of proteins bound to tags. These fusion proteins are labeled with compounds such as His-tags, biotin or antigens, which bind to the stationary phase specifically. After purification, some of these tags are usually removed and the pure protein is obtained.

Affinity chromatography often utilizes a biomolecule's affinity for a metal (Zn, Cu, Fe, etc.). Columns are often manually prepared. Traditional affinity columns are used as a preparative step to flush out unwanted biomolecules.

However, HPLC techniques exist that do utilize affinity chromatography properties. Immobilized Metal Affinity Chromatography (IMAC) is useful to separate aforementioned molecules based on the relative affinity for the metal (i.e. Dionex IMAC). Often these columns can be loaded with different metals to create a column with a targeted affinity.

Supercritical Fluid Chromatography

Supercritical fluid chromatography is a separation technique in which the mobile phase is a fluid above and relatively close to its critical temperature and pressure.

Techniques by Separation Mechanism

Ion Exchange Chromatography

Ion exchange chromatography (usually referred to as ion chromatography) uses an ion exchange mechanism to separate analytes based on their respective charges. It is usually performed in columns but can also be useful in planar mode. Ion exchange chromatography uses a charged stationary phase to separate charged compounds including anions, cations, amino acids, peptides, and proteins. In conventional methods the stationary phase is an ion exchange resin that carries charged functional groups that interact with oppositely charged groups of the compound to retain. There are two types of ion exchange chromatography: Cation-Exchange and Anion-Exchange. In the Cation-Exchange Chromatography the stationary phase has negative charge and the exchangeable ion is a cation, whereas, in the Anion-Exchange Chromatography the stationary phase has positive charge and the exchangeable ion is an anion. Ion exchange chromatography is commonly used to purify proteins using FPLC.

Size-exclusion Chromatography

Size-exclusion chromatography (SEC) is also known as *gel permeation chromatography* (GPC) or *gel filtration chromatography* and separates molecules according to their size (or more accurately according to their hydrodynamic diameter or hydrodynamic volume). Smaller molecules are able to enter the pores of the media and, therefore, molecules are trapped and removed from the flow of the mobile phase. The average residence time in the pores depends upon the effective size of the analyte molecules. However, molecules that are larger than the average pore size of the packing are excluded and thus suffer essentially no retention; such species are the first to be eluted. It is generally a low-resolution chromatography technique and thus it is often reserved for the final, "polishing" step of a purification. It is also useful for determining the tertiary structure and quaternary structure of purified proteins, especially since it can be carried out under native solution conditions.

Expanded Bed Adsorption Chromatographic Separation

An expanded bed chromatographic adsorption (EBA) column for a biochemical separation process comprises a pressure equalization liquid distributor having a self-cleaning function below a porous blocking sieve plate at the bottom of the expanded bed, an upper part nozzle assembly having a backflush cleaning function at the top of the expanded bed, a better distribution of the feedstock liquor added into the expanded bed ensuring that the fluid passed through the expanded bed layer displays a state of piston flow. The expanded bed layer displays a state of piston flow. The expanded bed chromatographic separation column has advantages of increasing the separation efficiency of the expanded bed.

Expanded-bed adsorption (EBA) chromatography is a convenient and effective technique for the capture of proteins directly from unclarified crude sample. In EBA chromatography, the settled bed is first expanded by upward flow of equilibration buffer. The crude feed, a mixture of soluble proteins, contaminants, cells, and cell debris, is then passed upward through the expanded bed. Target proteins are captured on the adsorbent, while particulates and contaminants pass through. A change to elution buffer while maintaining upward flow results in desorption of the target protein in expanded-bed mode. Alternatively, if the flow is reversed, the adsorbed particles will quickly settle and the proteins can be desorbed by an elution buffer. The mode used for elution (expanded-bed versus settled-bed) depends on the characteristics of the feed. After elution, the adsorbent is cleaned with a predefined cleaning-in-place (CIP) solution, with cleaning followed by either column regeneration (for further use) or storage.

Special Techniques

Reversed-phase Chromatography

Reversed-phase chromatography (RPC) is any liquid chromatography procedure in which the mobile phase is significantly more polar than the stationary phase. It is so named because in normal-phase liquid chromatography, the mobile phase is significantly less polar than the stationary phase. Hydrophobic molecules in the mobile phase tend to adsorb to the relatively hydrophobic stationary phase. Hydrophilic molecules in the mobile phase will tend to elute first. Separating columns typically comprise a C8 or C18 carbon-chain bonded to a silica particle substrate.

Hydrophobic Interaction Chromatography

Hydrophobic interactions between proteins and the chromatographic matrix can be exploited to purify proteins. In hydrophobic interaction chromatography the matrix material is lightly substituted with hydrophobic groups. These groups can range from methyl, ethyl, propyl, octyl, or phenyl groups. At high salt concentrations, non-polar sidechains on the surface on proteins "interact" with the hydrophobic groups; that is, both types of groups are excluded by the polar solvent (hydrophobic effects are augmented by increased ionic strength). Thus, the sample is applied to the column in a buffer which is highly polar. The eluant is typically an aqueous buffer with decreasing salt concentrations, increasing concentrations of detergent (which disrupts hydrophobic interactions), or changes in pH.

In general, Hydrophobic Interaction Chromatography (HIC) is advantageous if the sample is sensitive to pH change or harsh solvents typically used in other types of chromatography but not high salt concentrations. Commonly, it is the amount of salt in the buffer which is varied. In 2012, Müller and Franzreb described the effects of temperature on HIC using Bovine Serum Albumin (BSA) with four different types of hydrophobic resin. The study altered temperature as to effect the binding affinity of BSA onto the matrix. It was concluded that cycling temperature from 50 degrees to 10 degrees would not be adequate to effectively wash all BSA from the matrix but could be very effective if the column would only be used a few times. Using temperature to effect change allows labs to cut costs on buying salt and saves money.

If high salt concentrations along with temperature fluctuations want to be avoided you can use a more hydrophobic to compete with your sample to elute it. This so-called salt independent method of HIC showed a direct isolation of Human Immunoglobulin G (IgG) from serum with satisfactory yield and used Beta-cyclodextrin as a competitor to displace IgG from the matrix. This largely opens up the possibility of using HIC with samples which are salt sensitive as we know high salt concentrations precipitate proteins.

Two-dimensional Chromatography

In some cases, the chemistry within a given column can be insufficient to separate some analytes. It is possible to direct a series of unresolved peaks onto a second column with different physico-chemical (chemical classification) properties. Since the mechanism of retention on this new solid support is different from the first dimensional separation, it can be possible to separate compounds by two-dimensional chromatography that are indistinguishable by one-dimensional chromatography.

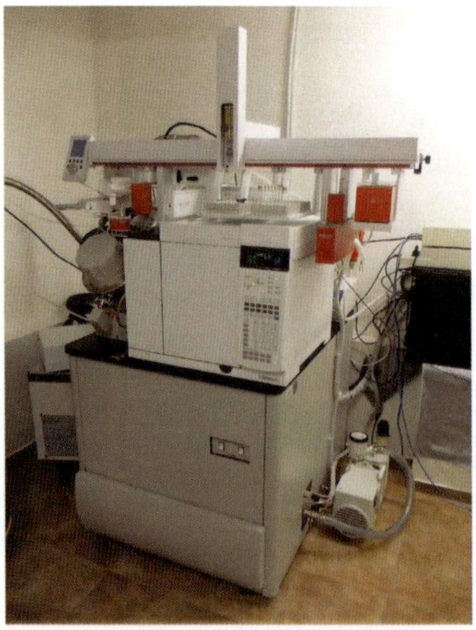

Two-dimensional chromatograph.

The sample is spotted at one corner of a square plate, developed, air-dried, then rotated by 90° and usually redeveloped in a second solvent system.

Simulated Moving-bed Chromatography

The simulated moving bed (SMB) technique is a variant of high performance liquid chromatography; it is used to separate particles and chemical compounds that would be difficult or impossible to resolve otherwise. This increased separation is brought about by a valve-and-column arrangement that is used to lengthen the stationary phase indefinitely. In the moving bed technique of preparative chromatography the feed entry and the analyte recovery are simultaneous and continuous, but because of practical difficulties with a continuously moving bed, simulated moving bed technique was proposed. In the simulated moving bed technique instead of moving the bed, the sample inlet and the analyte exit positions are moved continuously, giving the impression of a moving bed. True moving bed chromatography (TMBC) is only a theoretical concept. Its simulation, SMBC is achieved by the use of a multiplicity of columns in series and a complex valve arrangement, which provides for sample and solvent feed, and also analyte and waste takeoff at appropriate locations of any column, whereby it allows switching at regular intervals the sample entry in one direction, the solvent entry in the opposite direction, whilst changing the analyte and waste takeoff positions appropriately as well.

Pyrolysis Gas Chromatography

Pyrolysis–gas chromatography–mass spectrometry is a method of chemical analysis in which the sample is heated to decomposition to produce smaller molecules that are separated by gas chromatography and detected using mass spectrometry.

Pyrolysis is the thermal decomposition of materials in an inert atmosphere or a vacuum. The sample is put into direct contact with a platinum wire, or placed in a quartz sample tube, and rapidly heated to 600–1000 °C. Depending on the application even higher temperatures are used. Three different heating techniques are used in actual pyrolyzers: Isothermal furnace, inductive heating (Curie Point filament), and resistive heating using platinum filaments. Large molecules cleave at their weakest points and produce smaller, more volatile fragments. These fragments can be separated by gas chromatography. Pyrolysis GC chromatograms are typically complex because a wide range of different decomposition products is formed. The data can either be used as fingerprint to prove material identity or the GC/MS data is used to identify individual fragments to obtain structural information. To increase the volatility of polar fragments, various methylating reagents can be added to a sample before pyrolysis.

Besides the usage of dedicated pyrolyzers, pyrolysis GC of solid and liquid samples can be performed directly inside Programmable Temperature Vaporizer (PTV) injectors that provide quick heating (up to 30 °C/s) and high maximum temperatures of 600–650 °C. This is sufficient for some pyrolysis applications. The main advantage is that no dedicated instrument has to be purchased and pyrolysis can be performed as part of routine GC analysis. In this case quartz GC inlet liners have to be used. Quantitative data can be acquired, and good results of derivatization inside the PTV injector are published as well.

Fast Protein Liquid Chromatography

Fast protein liquid chromatography (FPLC), is a form of liquid chromatography that is often used to analyze or purify mixtures of proteins. As in other forms of chromatography, separation is possible because the different components of a mixture have different affinities for two materials, a moving fluid (the "mobile phase") and a porous solid (the stationary phase). In FPLC the mobile phase is an aqueous solution, or "buffer". The buffer flow rate is controlled by a positive-displacement pump and is normally kept constant, while the composition of the buffer can be varied by drawing fluids in different proportions from two or more external reservoirs. The stationary phase is a resin composed of beads, usually of cross-linked agarose, packed into a cylindrical glass or plastic column. FPLC resins are available in a wide range of bead sizes and surface ligands depending on the application.

Counter Current Chromatography

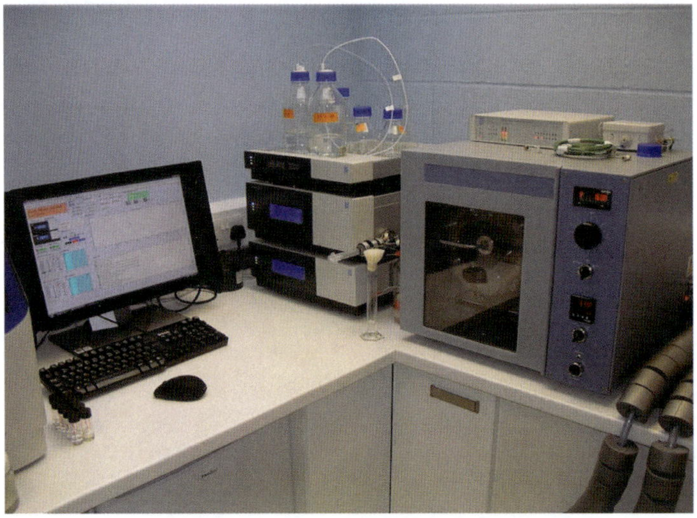

An example of a HPCCC system.

Counter current chromatography (CCC) is a type of liquid-liquid chromatography, where both the stationary and mobile phases are liquids. The operating principle of CCC equipment requires a column consisting of an open tube coiled around a bobbin. The bobbin is rotated in a double-axis gyratory motion (a cardioid), which causes a variable gravity (G) field to act on the column during each rotation. This motion causes the column to see one partitioning step per revolution and components of the sample separate in the column due to their partitioning coefficient between the two immiscible liquid phases used. There are many types of CCC available today. These include HSCCC (High Speed CCC) and HPCCC (High Performance CCC). HPCCC is the latest and best performing version of the instrumentation available currently.

Periodic Counter-current Chromatography

In contrast to Counter current chromatography, periodic counter-current chromatography (PCC) uses a solid stationary phase and only a liquid mobile phase. It thus is much more similar to conventional affinity chromatography than to counter current chromatography. PCC uses multiple columns, which during the loading phase are connected in line. This mode allows for overloading the first column in this series without losing product, which already breaks through the column before the resin is fully saturated. The breakthrough product is captured on the subsequent column(s). In a next step the columns are disconnected from one another. The first column is washed and eluted, while the other column(s) are still being loaded. Once the (initially) first column is re-equilibrated, it is re-introduced to the loading stream, but as last column. The process then continues in a cyclic fashion.

Chiral Chromatography

Chiral chromatography involves the separation of stereoisomers. In the case of enantiomers, these have no chemical or physical differences apart from being three-dimensional mirror images. Conventional chromatography or other separation processes are incapable of separating them. To enable chiral separations to take place, either the mobile phase or the stationary phase must themselves be made chiral, giving differing affinities between the analytes. Chiral chromatography HPLC columns (with a chiral stationary phase) in both normal and reversed phase are commercially available.

Aqueous Normal-phase Chromatography

Aqueous normal-phase (ANP) chromatography is characterized by the elution behavior of classical normal phase mode (i.e. where the mobile phase is significantly less polar than the stationary phase) in which water is one of the mobile phase solvent system components. It is distinguished from hydrophilic interaction liquid chromatography (HILIC) in that the retention mechanism is due to adsorption rather than partitioning.

Gas Chromatography

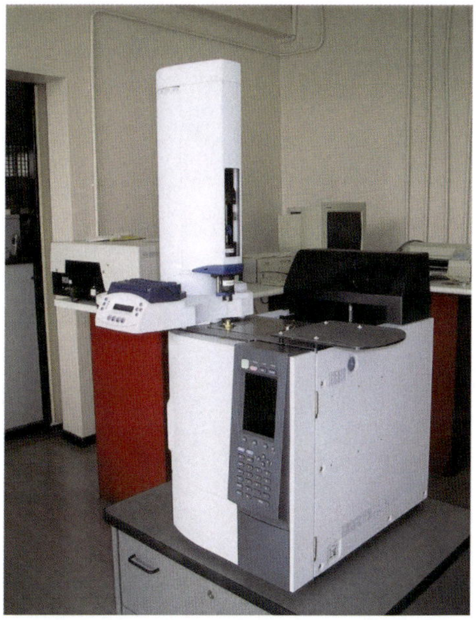

A gas chromatograph with a headspace sampler.

Gas chromatography (GC) is a common type of chromatography used in analytical chemistry for separating and analyzing compounds that can be vaporized without decomposition. Typical uses of GC include testing the purity of a particular substance, or separating the different components of a mixture (the relative amounts of such components can also be determined). In some situations, GC may help in identifying a compound. In preparative chromatography, GC can be used to prepare pure compounds from a mixture.

In gas chromatography, the *mobile phase* (or "moving phase") is a carrier gas, usually an inert gas such as helium or an unreactive gas such as nitrogen. Helium remains the most commonly used carrier gas in about 90% of instruments although hydrogen is preferred for improved separations. The *stationary phase* is a microscopic layer of liquid or polymer on an inert solid support, inside a piece of glass or metal tubing called a column (an homage to the fractionating column used in distillation). The instrument used to perform gas chromatography is called a *gas chromatograph* (or "aerograph", "gas separator").

The gaseous compounds being analyzed interact with the walls of the column, which is coated with a stationary phase. This causes each compound to elute at a different time, known as the *retention time* of the compound. The comparison of retention times is what gives GC its analytical usefulness.

Gas chromatography is in principle similar to column chromatography (as well as other forms of chromatography, such as HPLC, TLC), but has several notable differences. First, the process of separating the compounds in a mixture is carried out between a liquid stationary phase and a gas mobile phase, whereas in column chromatography the stationary phase is a solid and the mobile phase is a liquid. Hence the full name of the procedure is "Gas–liquid chromatography", referring to the mobile and stationary phases, respectively. Second, the column through which the gas phase passes is located in an oven where the temperature of the gas can be controlled, whereas column chromatography (typically) has no such temperature control. Finally, the concentration of a compound in the gas phase is solely a function of the vapor pressure of the gas.

Gas chromatography is also sometimes known as vapor-phase chromatography (VPC), or gas–liquid partition chromatography (GLPC). These alternative names, as well as their respective abbreviations, are frequently used in scientific literature. Strictly speaking, GLPC is the most correct terminology, and is thus preferred by many authors.

GC Analysis

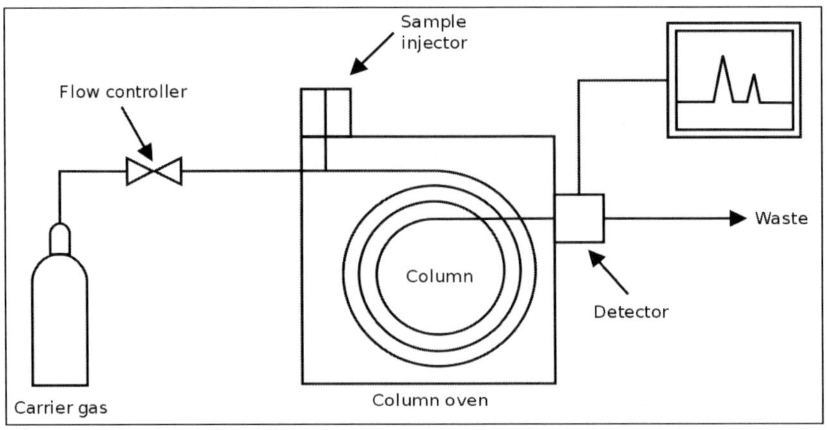

Diagram of a gas chromatograph.

A gas chromatograph is a chemical analysis instrument for separating chemicals in a complex sample. A gas chromatograph uses a flow-through narrow tube known as the *column*, through which different chemical constituents of a sample pass in a gas stream (carrier gas, *mobile phase*) at different rates depending on their various chemical and physical properties and their interaction with a specific column filling, called the *stationary phase*. As the chemicals exit the end of the column, they are detected and identified electronically. The function of the stationary phase in the column is to separate different components, causing each one to exit the column at a different time (*retention time*). Other parameters that can be used to alter the order or time of retention are the carrier gas flow rate, column length and the temperature.

In a GC analysis, a known volume of gaseous or liquid analyte is injected into the "entrance" (head) of the column, usually using a microsyringe (or, solid phase microextraction fibers, or a gas source switching system). As the carrier gas sweeps the analyte molecules through the column, this motion is inhibited by the adsorption of the analyte molecules either onto the column walls or onto packing materials in the column. The rate at which the molecules progress along the column depends on the strength of adsorption, which in turn depends on the type of molecule and on the stationary phase materials. Since each type of molecule has a different rate of progression, the various components of the analyte mixture are separated as they progress along the column and reach the end of the column at different times (retention time). A detector is used to monitor the outlet stream from the column; thus, the time at which each component reaches the outlet and the amount of that component can be determined. Generally, substances are identified (qualitatively) by the order in which they emerge (elute) from the column and by the retention time of the analyte in the column.

Physical Components

Autosamplers

The autosampler provides the means to introduce a sample automatically into the inlets. Manual insertion of the sample is possible but is no longer common. Automatic insertion provides better reproducibility and time-optimization.

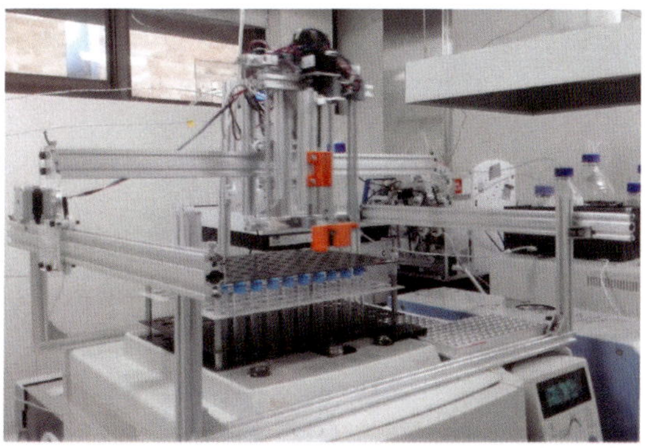

An autosampler for liquid or gaseous samples based on a microsyringe.

Different kinds of autosamplers exist. Autosamplers can be classified in relation to sample capacity (auto-injectors vs. autosamplers, where auto-injectors can work a small number of samples), to robotic technologies (XYZ robot vs. rotating robot – the most common), or to analysis:

- Liquid.
- Static head-space by syringe technology.
- Dynamic head-space by transfer-line technology.
- Solid phase microextraction (SPME).

Inlets

The column inlet (or injector) provides the means to introduce a sample into a continuous flow of carrier gas. The inlet is a piece of hardware attached to the column head.

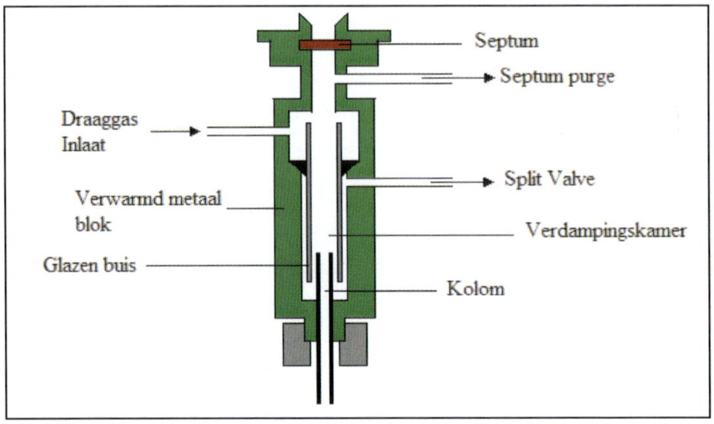

Split/splitless inlet.

Common inlet types are:

- S/SL (split/splitless) injector: A sample is introduced into a heated small chamber via a syringe through a septum – the heat facilitates volatilization of the sample and sample matrix. The carrier gas then either sweeps the entirety (splitless mode) or a portion (split mode) of the sample into the column. In split mode, a part of the sample/carrier gas mixture in the injection chamber is exhausted through the split vent. Split injection is preferred when working with samples with high analyte concentrations (>0.1%) whereas splitless injection is best suited for trace analysis with low amounts of analytes (<0.01%). In splitless mode the split valve opens after a pre-set amount of time to purge heavier elements that would otherwise contaminate the system. This pre-set (splitless) time should be optimized, the shorter time (e.g., 0.2 min) ensures less tailing but loss in response, the longer time (2 min) increases tailing but also signal.

- On-column inlet: The sample is here introduced directly into the column in its entirety without heat, or at a temperature below the boiling point of the solvent. The low temperature condenses the sample into a narrow zone. The column and inlet can then be heated, releasing the sample into the gas phase. This ensures the lowest possible temperature for chromatography and keeps samples from decomposing above their boiling point.

- PTV injector: Temperature-programmed sample introduction was first described by Vogt in 1979. Originally Vogt developed the technique as a method for the introduction of large sample volumes (up to 250 μL) in capillary GC. Vogt introduced the sample into the liner at a controlled injection rate. The temperature of the liner was chosen slightly below the boiling point of the solvent. The low-boiling solvent was continuously evaporated and vented through the split line. Based on this technique, Poy developed the programmed temperature vaporising injector; PTV. By introducing the sample at a low initial liner temperature many of the disadvantages of the classic hot injection techniques could be circumvented.

- Gas source inlet or gas switching valve: Gaseous samples in collection bottles are connected to what is most commonly a six-port switching valve. The carrier gas flow is not interrupted while a sample can be expanded into a previously evacuated sample loop. Upon switching, the contents of the sample loop are inserted into the carrier gas stream.

- P/T (Purge-and-Trap) system: An inert gas is bubbled through an aqueous sample causing insoluble volatile chemicals to be purged from the matrix. The volatiles are 'trapped' on an absorbent column (known as a trap or concentrator) at ambient temperature. The trap is then heated and the volatiles are directed into the carrier gas stream. Samples requiring preconcentration or purification can be introduced via such a system, usually hooked up to the S/SL port.

The choice of carrier gas (mobile phase) is important. Hydrogen has a range of flow rates that are comparable to helium in efficiency. However, helium may be more efficient and provide the best separation if flow rates are optimized. Helium is non-flammable and works with a greater number of detectors and older instruments. Therefore, helium is the most common carrier gas used. However, the price of helium has gone up considerably over recent years, causing an increasing number of chromatographers to switch to hydrogen gas. Historical use, rather than rational consideration, may contribute to the continued preferential use of helium.

Detectors

The most commonly used detectors are the flame ionization detector (FID) and the thermal conductivity detector (TCD). Both are sensitive to a wide range of components, and both work over a wide range of concentrations. While TCDs are essentially universal and can be used to detect any component other than the carrier gas (as long as their thermal conductivities are different from that of the carrier gas, at detector temperature). FIDs are sensitive primarily to hydrocarbons, and are more sensitive to them than TCD. However, a FID cannot detect water. Both detectors are also quite robust. Since TCD is non-destructive, it can be operated in-series before a FID (destructive), thus providing complementary detection of the same analytes. Other detectors are sensitive only to specific types of substances, or work well only in narrower ranges of concentrations.

Thermal conductivity detector (TCD) relies on the thermal conductivity of matter passing around a tungsten -rhenium filament with a current traveling through it. In this set up helium or nitrogen serve as the carrier gas because of their relatively high thermal conductivity which keep the filament cool and maintain uniform resistivity and electrical efficiency of the filament. However, when analyte molecules elute from the column, mixed with carrier gas, the thermal conductivity decreases and this causes a detector response. The response is due to the decreased thermal conductivity causing an increase in filament temperature and resistivity resulting in fluctuations in voltage. Detector sensitivity is proportional to filament current while it is inversely proportional to the immediate environmental temperature of that detector as well as flow rate of the carrier gas.

In a flame ionization detector (FID), electrodes are placed adjacent to a flame fueled by hydrogen/air near the exit of the column, and when carbon containing compounds exit the column they are pyrolyzed by the flame. This detector works only for organic/hydrocarbon containing compounds due to the ability of the carbons to form cations and electrons upon pyrolysis which generates a current between the electrodes. The increase in current is translated and appears as a peak in a chromatogram. FIDs have low detection limits (a few picograms per second) but they are unable to generate ions from carbonyl containing carbons. FID compatible carrier gasses include helium, hydrogen, nitrogen, and argon.

Alkali flame detector (AFD) or alkali flame ionization detector (AFID) has high sensitivity to nitrogen and phosphorus, similar to NPD. However, the alkaline metal ions are supplied with the hydrogen gas, rather than a bead above the flame. For this reason AFD does not suffer the "fatigue" of the NPD, but provides a constant sensitivity over long period of time. In addition, when alkali ions are not added to the flame, AFD operates like a standard FID. A catalytic combustion detector (CCD) measures combustible hydrocarbons and hydrogen. Discharge ionization detector (DID) uses a high-voltage electric discharge to produce ions.

The polyarc reactor is an add-on to new or existing GC-FID instruments that converts all organic compounds to methane molecules prior to their detection by the FID. This technique can be used to improve the response of the FID and allow for the detection of many more carbon-containing compounds. The complete conversion of compounds to methane and the now equivalent response in the detector also eliminates the need for calibrations and standards because response factors are all equivalent to those of methane. This allows for the rapid analysis of complex mixtures that contain molecules where standards are not available.

Flame photometric detector (FPD) uses a photomultiplier tube to detect spectral lines of the compounds as they are burned in a flame. Compounds eluting off the column are carried into a hydrogen fueled flame which excites specific elements in the molecules, and the excited elements (P, S, Halogens, etc.) emit light of specific characteristic wavelengths. The emitted light is filtered and detected by a photomultiplier tube. In particular, phosphorus emission is around 510–536 nm and sulfur emission is at 394 nm. With an atomic emission detector (AED), a sample eluting from a column enters a chamber which is energized by microwaves that induce a plasma. The plasma causes the analyte sample to decompose and certain elements generate an atomic emission spectra. The atomic emission spectra is diffracted by a diffraction grating and detected by a series of photomultiplier tubes or photo diodes.

Electron capture detector (ECD) uses a radioactive beta particle (electron) source to measure the degree of electron capture. ECD are used for the detection of molecules containing electronegative/withdrawing elements and functional groups like halogens, carbonyl, nitriles, nitro groups, and organometalics. In this type of detector either nitrogen or 5% methane in argon is used as the mobile phase carrier gas. The carrier gas passes between two electrodes placed at the end of the column, and adjacent to the anode (negative electrode) resides a radioactive foil such as 63Ni. The radioactive foil emits a beta particle (electron) which collides with and ionizes the carrier gas to generate more ions resulting in a current. When analyte molecules with electronegative/withdrawing elements or functional groups electrons are captured which results in a decrease in current generating a detector response.

Nitrogen–phosphorus detector (NPD), a form of thermionic detector where nitrogen and phosphorus alter the work function on a specially coated bead and a resulting current is measured.

Dry electrolytic conductivity detector (DELCD) uses an air phase and high temperature (v. Coulsen) to measure chlorinated compounds.

Mass spectrometer (MS), also called GC-MS; highly effective and sensitive, even in a small quantity of sample. This detector can be used to identify the analytes in chromatograms by their mass spectrum. Some GC-MS are connected to an NMR spectrometer which acts as a backup detector. This combination is known as GC-MS-NMR. Some GC-MS-NMR are connected to an infrared spectrophotometer which acts as a backup detector. This combination is known as GC-MS-NMR-IR. It must, however, be stressed this is very rare as most analyses needed can be concluded via purely GC-MS.

Vacuum ultraviolet (VUV) represents the most recent development in gas chromatography detectors. Most chemical species absorb and have unique gas phase absorption cross sections in the approximately 120–240 nm VUV wavelength range monitored. Where absorption cross sections are known for analytes, the VUV detector is capable of absolute determination (without calibration) of the number of molecules present in the flow cell in the absence of chemical interferences.

Other detectors include the Hall electrolytic conductivity detector (ElCD), helium ionization detector (HID), infrared detector (IRD), photo-ionization detector (PID), pulsed discharge ionization detector (PDD), and thermionic ionization detector (TID).

Methods

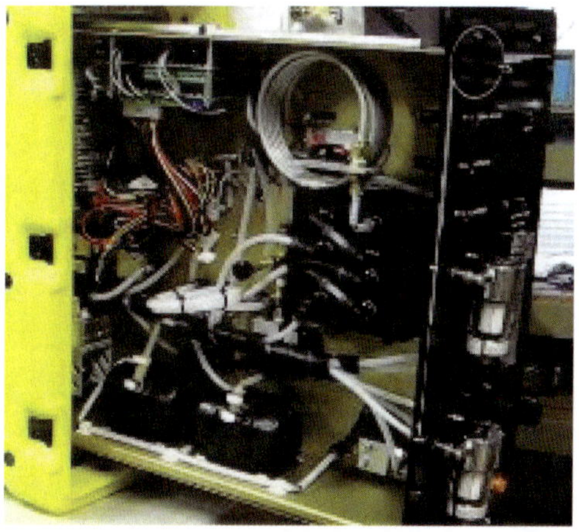

This image above shows the interior of a Eclipse Gas Chromatograph that runs continuously in three-minute cycles. Two valves are used to switch the test gas into the sample loop. After filling the sample loop with test gas, the valves are switched again applying carrier gas pressure to the sample loop and forcing the sample through the column for separation.

The method is the collection of conditions in which the GC operates for a given analysis. Method development is the process of determining what conditions are adequate and ideal for the analysis required.

Conditions which can be varied to accommodate a required analysis include inlet temperature, detector temperature, column temperature and temperature program, carrier gas and carrier gas flow rates, the column's stationary phase, diameter and length, inlet type and flow rates, sample size and injection technique. Depending on the detector(s) installed on the GC, there may be a number of detector conditions that can also be varied. Some GCs also include valves which can change the route of sample and carrier flow. The timing of the opening and closing of these valves can be important to method development.

Carrier Gas Selection and Flow Rates

Typical carrier gases include helium, nitrogen, argon, hydrogen and air. Which gas to use is usually determined by the detector being used, for example, a DID requires helium as the carrier gas.

When analyzing gas samples, however, the carrier is sometimes selected based on the sample's matrix, for example, when analyzing a mixture in argon, an argon carrier is preferred, because the argon in the sample does not show up on the chromatogram. Safety and availability can also influence carrier selection, for example, hydrogen is flammable, and high-purity helium can be difficult to obtain in some areas of the world. As a result of helium becoming more scarce, hydrogen is often being substituted for helium as a carrier gas in several applications.

The purity of the carrier gas is also frequently determined by the detector, though the level of sensitivity needed can also play a significant role. Typically, purities of 99.995% or higher are used. The most common purity grades required by modern instruments for the majority of sensitivities are 5.0 grades, or 99.999% pure meaning that there is a total of 10 ppm of impurities in the carrier gas that could affect the results. The highest purity grades in common use are 6.0 grades, but the need for detection at very low levels in some forensic and environmental applications has driven the need for carrier gases at 7.0 grade purity and these are now commercially available. Trade names for typical purities include "Zero Grade," "Ultra-High Purity (UHP) Grade," "4.5 Grade" and "5.0 Grade."

The carrier gas linear velocity affects the analysis in the same way that temperature does. The higher the linear velocity the faster the analysis, but the lower the separation between analytes. Selecting the linear velocity is therefore the same compromise between the level of separation and length of analysis as selecting the column temperature. The linear velocity will be implemented by means of the carrier gas flow rate, with regards to the inner diameter of the column.

With GCs made before the 1990s, carrier flow rate was controlled indirectly by controlling the carrier inlet pressure, or "column head pressure." The actual flow rate was measured at the outlet of the column or the detector with an electronic flow meter, or a bubble flow meter, and could be an involved, time consuming, and frustrating process. It was not possible to vary the pressure setting during the run, and thus the flow was essentially constant during the analysis. The relation between flow rate and inlet pressure is calculated with Poiseuille's equation for compressible fluids.

Many modern GCs, however, electronically measure the flow rate, and electronically control the carrier gas pressure to set the flow rate. Consequently, carrier pressures and flow rates can be adjusted during the run, creating pressure/flow programs similar to temperature programs.

Stationary Compound Selection

The polarity of the solute is crucial for the choice of stationary compound, which in an optimal case would have a similar polarity as the solute. Common stationary phases in open tubular columns are cyanopropylphenyl dimethyl polysiloxane, carbowax polyethyleneglycol, biscyanopropyl cyanopropylphenyl polysiloxane and diphenyl dimethyl polysiloxane. For packed columns more options are available.

Inlet Types and Flow Rates

The choice of inlet type and injection technique depends on if the sample is in liquid, gas, adsorbed, or solid form, and on whether a solvent matrix is present that has to be vaporized. Dissolved samples can be introduced directly onto the column via a COC injector, if the conditions are well known; if a solvent matrix has to be vaporized and partially removed, a S/SL injector is used (most common injection technique); gaseous samples (e.g., air cylinders) are usually injected using a gas switching valve system; adsorbed samples (e.g., on adsorbent tubes) are introduced using either an external (on-line or off-line) desorption apparatus such as a purge-and-trap system, or are desorbed in the injector (SPME applications).

Sample Size and Injection Technique

Sample Injection

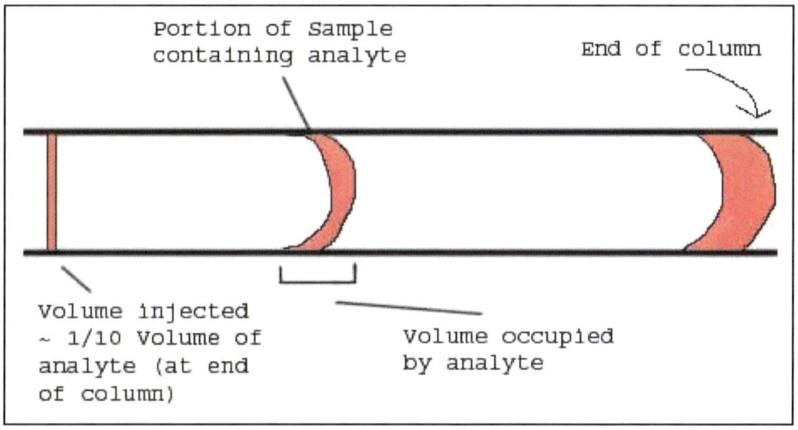

The rule of ten in gas chromatography.

The real chromatographic analysis starts with the introduction of the sample onto the column. The development of capillary gas chromatography resulted in many practical problems with the injection technique. The technique of on-column injection, often used with packed columns, is usually not possible with capillary columns. In the injection system in the capillary gas chromatograph the amount injected should not overload the column and the width of the injected plug should be small compared to the spreading due to the chromatographic process. Failure to comply with this latter requirement will reduce the separation capability of the column. As a general rule, the volume injected, V_{inj}, and the volume of the detector cell, V_{det}, should be about 1/10 of the volume occupied by the portion of sample containing the molecules of interest (analytes) when they exit the column.

Some general requirements which a good injection technique should fulfill are that it should be possible to obtain the column's optimum separation efficiency, it should allow accurate and reproducible injections of small amounts of representative samples, it should induce no change in sample composition, it should not exhibit discrimination based on differences in boiling point, polarity, concentration or thermal/catalytic stability, and it should be applicable for trace analysis as well as for undiluted samples.

However, there are a number of problems inherent in the use of syringes for injection. Even the best syringes claim an accuracy of only 3%, and in unskilled hands, errors are much larger. The needle may cut small pieces of rubber from the septum as it injects sample through it. These can block the needle and prevent the syringe filling the next time it is used. It may not be obvious that this has happened. A fraction of the sample may get trapped in the rubber, to be released during subsequent injections. This can give rise to ghost peaks in the chromatogram.

There may be selective loss of the more volatile components of the sample by evaporation from the tip of the needle.

Column Selection

The choice of column depends on the sample and the active measured. The main chemical attribute regarded when choosing a column is the polarity of the mixture, but functional groups can play a large part in column selection. The polarity of the sample must closely match the polarity of the column stationary phase to increase resolution and separation while reducing run time. The separation and run time also depends on the film thickness (of the stationary phase), the column diameter and the column length.

Column Temperature and Temperature Program

A gas chromatography oven, open to show a capillary column.

The column(s) in a GC are contained in an oven, the temperature of which is precisely controlled electronically. (When discussing the "temperature of the column," an analyst is technically referring to the temperature of the column oven).

The rate at which a sample passes through the column is directly proportional to the temperature of the column. The higher the column temperature, the faster the sample moves through the column. However, the faster a sample moves through the column, the less it interacts with the stationary phase, and the less the analytes are separated.

In general, the column temperature is selected to compromise between the length of the analysis and the level of separation.

A method which holds the column at the same temperature for the entire analysis is called "isothermal." Most methods, however, increase the column temperature during the analysis, the initial temperature, rate of temperature increase (the temperature "ramp"), and final temperature are called the temperature program.

A temperature program allows analytes that elute early in the analysis to separate adequately, while shortening the time it takes for late-eluting analytes to pass through the column.

Data Reduction and Analysis

Qualitative Analysis

Generally, chromatographic data is presented as a graph of detector response (y-axis) against retention time (x-axis), which is called a chromatogram. This provides a spectrum of peaks for a sample representing the analytes present in a sample eluting from the column at different times. Retention time can be used to identify analytes if the method conditions are constant. Also, the pattern of peaks will be constant for a sample under constant conditions and can identify complex mixtures of analytes. However, in most modern applications, the GC is connected to a mass spectrometer or similar detector that is capable of identifying the analytes represented by the peaks.

Quantitative Analysis

The area under a peak is proportional to the amount of analyte present in the chromatogram. By calculating the area of the peak using the mathematical function of integration, the concentration of an analyte in the original sample can be determined. Concentration can be calculated using a calibration curve created by finding the response for a series of concentrations of analyte, or by determining the relative response factor of an analyte. The relative response factor is the expected ratio of an analyte to an internal standard (or external standard) and is calculated by finding the response of a known amount of analyte and a constant amount of internal standard (a chemical added to the sample at a constant concentration, with a distinct retention time to the analyte).

In most modern GC-MS systems, computer software is used to draw and integrate peaks, and match MS spectra to library spectra.

Applications

In general, substances that vaporize below 300 °C (and therefore are stable up to that temperature) can be measured quantitatively. The samples are also required to be salt-free; they should not contain ions. Very minute amounts of a substance can be measured, but it is often required that the sample must be measured in comparison to a sample containing the pure, suspected substance known as a reference standard.

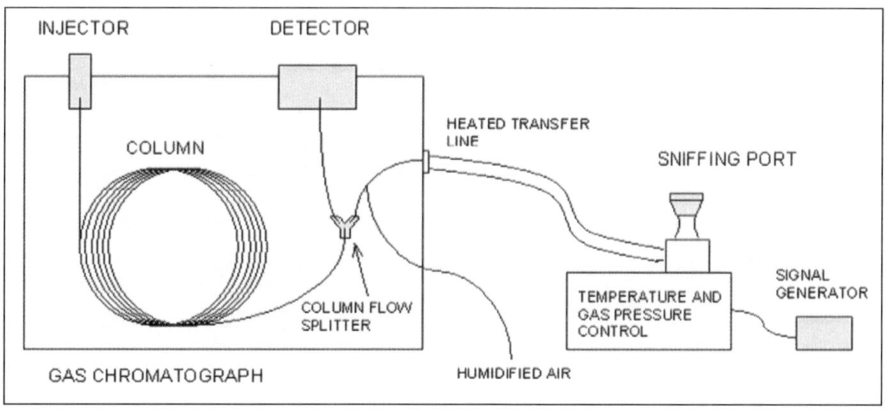

Various temperature programs can be used to make the readings more meaningful; for example to differentiate between substances that behave similarly during the GC process.

Professionals working with GC analyze the content of a chemical product, for example in assuring the quality of products in the chemical industry; or measuring toxic substances in soil, air or water. GC is very accurate if used properly and can measure picomoles of a substance in a 1 ml liquid sample, or parts-per-billion concentrations in gaseous samples.

In practical courses at colleges, students sometimes get acquainted to the GC by studying the contents of Lavender oil or measuring the ethylene that is secreted by *Nicotiana benthamiana* plants after artificially injuring their leaves. These GC analyse hydrocarbons (C2-C40+). In a typical experiment, a packed column is used to separate the light gases, which are then detected with a TCD. The hydrocarbons are separated using a capillary column and detected with a FID. A complication with light gas analyses that include H_2 is that He, which is the most common and most sensitive inert carrier (sensitivity is proportional to molecular mass) has an almost identical thermal conductivity to hydrogen (it is the difference in thermal conductivity between two separate filaments in a Wheatstone Bridge type arrangement that shows when a component has been eluted). For this reason, dual TCD instruments used with a separate channel for hydrogen that uses nitrogen as a carrier are common. Argon is often used when analysing gas phase chemistry reactions such as F-T synthesis so that a single carrier gas can be used rather than two separate ones. The sensitivity is reduced, but this is a trade off for simplicity in the gas supply.

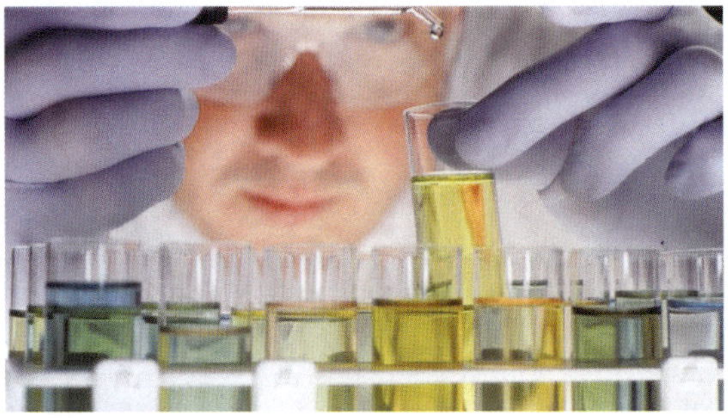

Gas Chromatography is used extensively in forensic science. Disciplines as diverse as solid drug dose (pre-consumption form) identification and quantification, arson investigation, paint chip analysis, and toxicology cases, employ GC to identify and quantify various biological specimens and crime-scene evidence.

High-performance Liquid Chromatography

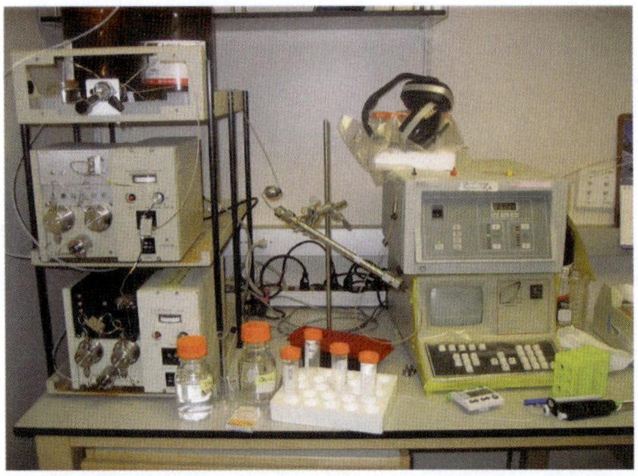

An HPLC. From left to right: A pumping device generating a gradient of two different solvents- a steel-enforced column and a detector for measuring the absorbance.

High-performance liquid chromatography (HPLC; formerly referred to as high-pressure liquid chromatography) is a technique in analytical chemistry used to separate, identify, and quantify each component in a mixture. It relies on pumps to pass a pressurized liquid solvent containing the sample mixture through a column filled with a solid adsorbent material. Each component in the sample interacts slightly differently with the adsorbent material, causing different flow rates for the different components and leading to the separation of the components as they flow out of the column.

A modern self-contained HPLC.

HPLC has been used for manufacturing (e.g., during the production process of pharmaceutical and biological products), legal (e.g., detecting performance enhancement drugs in urine), research (e.g., separating the components of a complex biological sample, or of similar synthetic chemicals from each other), and medical (e.g., detecting vitamin D levels in blood serum) purposes.

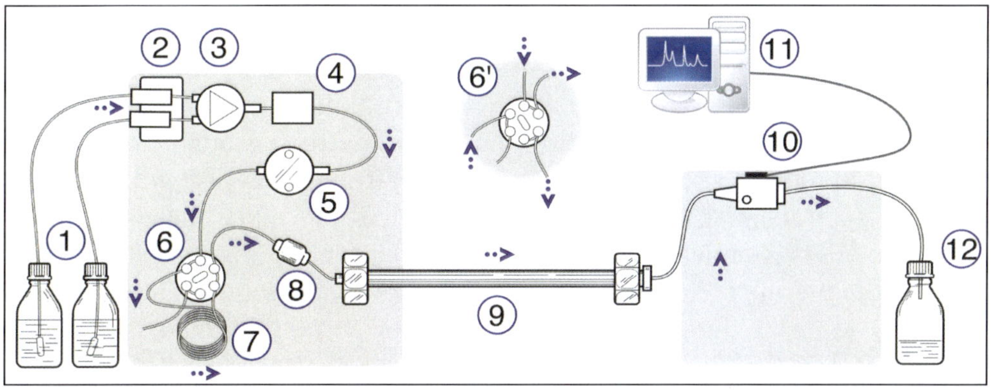

Schematic representation of an HPLC unit. (1) Solvent reservoirs, (2) Solvent degasser, (3) Gradient valve, (4) Mixing vessel for delivery of the mobile phase, (5) High-pressure pump, (6) Switching valve in "inject position", (6') Switching valve in "load position", (7) Sample injection loop, (8) Pre-column (guard column), (9) Analytical column, (10) Detector (*i.e.*, IR, UV), (11) Data acquisition, (12) Waste or fraction collector.

Chromatography can be described as a mass transfer process involving adsorption. HPLC relies on pumps to pass a pressurized liquid and a sample mixture through a column filled with adsorbent, leading to the separation of the sample components. The active component of the column, the adsorbent, is typically a granular material made of solid particles (*e.g.*, silica, polymers, etc.), 2–50 µm in size. The components of the sample mixture are separated from each other due to their different degrees of interaction with the adsorbent particles. The pressurized liquid is typically a mixture of solvents (*e.g.*, water, acetonitrile and methanol) and is referred to as a "mobile phase". Its composition and temperature play a major role in the separation process by influencing the interactions taking place between sample components and adsorbent. These interactions are physical in nature, such as hydrophobic (dispersive), dipole–dipole and ionic, most often a combination.

HPLC is distinguished from traditional ("low pressure") liquid chromatography because operational pressures are significantly higher (50–350 bar), while ordinary liquid chromatography typically relies on the force of gravity to pass the mobile phase through the column. Due to the small sample amount separated in analytical HPLC, typical column dimensions are 2.1–4.6 mm diameter, and 30–250 mm length. Also HPLC columns are made with smaller adsorbent particles (2–50 µm in average particle size). This gives HPLC superior resolving power (the ability to distinguish between compounds) when separating mixtures, which makes it a popular chromatographic technique.

The schematic of a HPLC instrument typically includes a degasser, sampler, pumps, and a detector. The sampler brings the sample mixture into the mobile phase stream which carries it into the column. The pumps deliver the desired flow and composition of the mobile phase through the column. The detector generates a signal proportional to the amount of sample component emerging from the column, hence allowing for quantitative analysis of the sample components. A digital microprocessor and user software control the HPLC instrument and provide data analysis. Some models of mechanical pumps in a HPLC instrument can mix multiple solvents together in ratios changing in time, generating a composition gradient in the mobile phase. Various detectors are in common use, such as UV/Vis, photodiode array (PDA) or based on mass spectrometry. Most HPLC instruments also have a column oven that allows for adjusting the temperature at which the separation is performed.

Operation

The sample mixture to be separated and analyzed is introduced, in a discrete small volume (typically microliters), into the stream of mobile phase percolating through the column. The components of the sample move through the column at different velocities, which are a function of specific physical interactions with the adsorbent (also called stationary phase). The velocity of each component depends on its chemical nature, on the nature of the stationary phase (column) and on the composition of the mobile phase. The time at which a specific analyte elutes (emerges from the column) is called its retention time. The retention time measured under particular conditions is an identifying characteristic of a given analyte.

Many different types of columns are available, filled with adsorbents varying in particle size, and in the nature of their surface ("surface chemistry"). The use of smaller particle size packing materials requires the use of higher operational pressure ("backpressure") and typically improves chromatographic resolution (the degree of peak separation between consecutive analytes emerging from the column). Sorbent particles may be hydrophobic or polar in nature.

Common mobile phases used include any miscible combination of water with various organic solvents (the most common are acetonitrile and methanol). Some HPLC techniques use water-free mobile phases. The aqueous component of the mobile phase may contain acids (such as formic, phosphoric or trifluoroacetic acid) or salts to assist in the separation of the sample components. The composition of the mobile phase may be kept constant ("isocratic elution mode") or varied ("gradient elution mode") during the chromatographic analysis. Isocratic elution is typically effective in the separation of sample components that are very different in their affinity for the stationary phase. In gradient elution the composition of the mobile phase is varied typically from low to high eluting strength. The eluting strength of the mobile phase is reflected by analyte retention times with high eluting strength producing fast elution (short retention times). A typical gradient profile in reversed phase chromatography might start at 5% acetonitrile (in water or aqueous buffer) and progress linearly to 95% acetonitrile over 5–25 minutes. Periods of constant mobile phase composition may be part of any gradient profile. For example, the mobile phase composition may be kept constant at 5% acetonitrile for 1–3 min, followed by a linear change up to 95% acetonitrile.

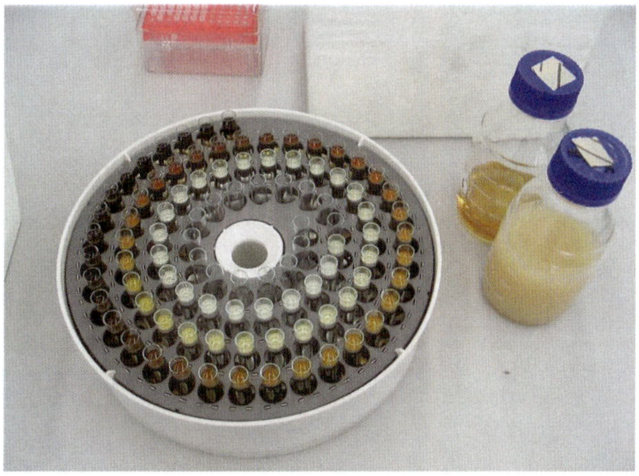

A rotary fraction collector collecting HPLC output. The system is being used to isolate a fraction containing Complex I from *E. coli* plasma membranes. About 50 litres of bacteria were needed to isolate this amount.

The chosen composition of the mobile phase (also called eluent) depends on the intensity of interactions between various sample components ("analytes") and stationary phase (e.g., hydrophobic interactions in reversed-phase HPLC). Depending on their affinity for the stationary and mobile phases analytes partition between the two during the separation process taking place in the column. This partitioning process is similar to that which occurs during a liquid–liquid extraction but is continuous, not step-wise. In this example, using a water/acetonitrile gradient, more hydrophobic components will elute (come off the column) late, once the mobile phase gets more concentrated in acetonitrile (i.e., in a mobile phase of higher eluting strength).

The choice of mobile phase components, additives (such as salts or acids) and gradient conditions depends on the nature of the column and sample components. Often a series of trial runs is performed with the sample in order to find the HPLC method which gives adequate separation.

Types

Partition Chromatography

Partition chromatography was one of the first kinds of chromatography that chemists developed. The partition coefficient principle has been applied in paper chromatography, thin layer chromatography, gas phase and liquid–liquid separation applications. The 1952 Nobel Prize in chemistry was earned by Archer John Porter Martin and Richard Laurence Millington Synge for their development of the technique, which was used for their separation of amino acids. Partition chromatography uses a retained solvent, on the surface or within the grains or fibers of an "inert" solid supporting matrix as with paper chromatography; or takes advantage of some coulombic and hydrogen donor interaction with the stationary phase. Analyte molecules partition between a liquid stationary phase and the eluent. Just as in Hydrophilic Interaction Chromatography (HILIC; a sub-technique within HPLC), this method separates analytes based on differences in their polarity. HILIC most often uses a bonded polar stationary phase and a mobile phase made primarily of acetonitrile with water as the strong component. Partition HPLC has been used historically on unbonded silica or alumina supports. Each works effectively for separating analytes by relative polar differences. HILIC bonded phases have the advantage of separating acidic, basic and neutral solutes in a single chromatographic run.

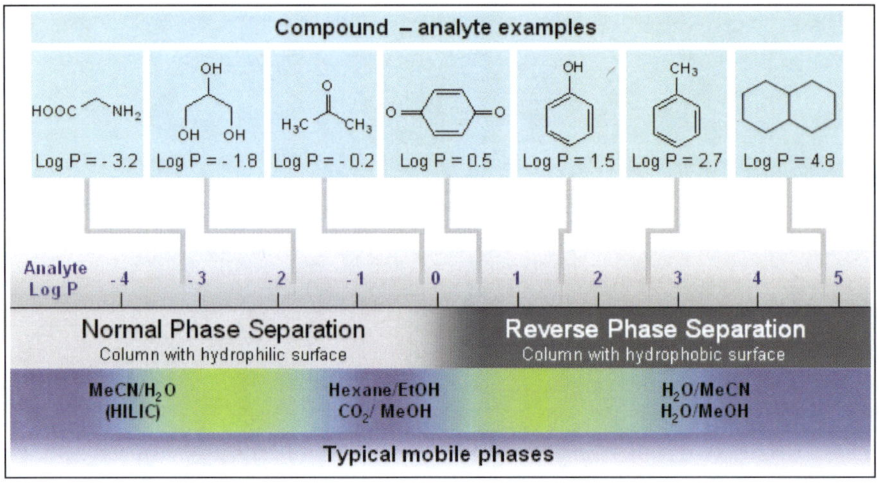

HILIC Partition Technique Useful Range.

The polar analytes diffuse into a stationary water layer associated with the polar stationary phase and are thus retained. The stronger the interactions between the polar analyte and the polar stationary phase (relative to the mobile phase) the longer the elution time. The interaction strength depends on the functional groups part of the analyte molecular structure, with more polarized groups (e.g., hydroxyl-) and groups capable of hydrogen bonding inducing more retention. Coulombic (electrostatic) interactions can also increase retention. Use of more polar solvents in the mobile phase will decrease the retention time of the analytes, whereas more hydrophobic solvents tend to increase retention times.

Normal–phase Chromatography

Normal–phase chromatography was one of the first kinds of HPLC that chemists developed. Also known as normal-phase HPLC (NP-HPLC) this method separates analytes based on their affinity for a polar stationary surface such as silica, hence it is based on analyte ability to engage in polar interactions (such as hydrogen-bonding or dipole-dipole type of interactions) with the sorbent surface. NP-HPLC uses a non-polar, non-aqueous mobile phase (e.g., Chloroform), and works effectively for separating analytes readily soluble in non-polar solvents. The analyte associates with and is retained by the polar stationary phase. Adsorption strengths increase with increased analyte polarity. The interaction strength depends not only on the functional groups present in the structure of the analyte molecule, but also on steric factors. The effect of steric hindrance on interaction strength allows this method to resolve (separate) structural isomers.

The use of more polar solvents in the mobile phase will decrease the retention time of analytes, whereas more hydrophobic solvents tend to induce slower elution (increased retention times). Very polar solvents such as traces of water in the mobile phase tend to adsorb to the solid surface of the stationary phase forming a stationary bound (water) layer which is considered to play an active role in retention. This behavior is somewhat peculiar to normal phase chromatography because it is governed almost exclusively by an adsorptive mechanism (*i.e.*, analytes interact with a solid surface rather than with the solvated layer of a ligand attached to the sorbent surface). Adsorption chromatography is still widely used for structural isomer separations in both column and thin-layer chromatography formats on activated (dried) silica or alumina supports.

Partition- and NP-HPLC fell out of favor in the 1970s with the development of reversed-phase HPLC because of poor reproducibility of retention times due to the presence of a water or protic organic solvent layer on the surface of the silica or alumina chromatographic media. This layer changes with any changes in the composition of the mobile phase (e.g., moisture level) causing drifting retention times.

Recently, partition chromatography has become popular again with the development of Hilic bonded phases which demonstrate improved reproducibility, and due to a better understanding of the range of usefulness of the technique.

Displacement Chromatography

The basic principle of displacement chromatography is: A molecule with a high affinity for the chromatography matrix (the displacer) will compete effectively for binding sites, and thus displace all molecules with lesser affinities. There are distinct differences between displacement and elution chromatography. In elution mode, substances typically emerge from a column in narrow, Gaussian peaks. Wide separation of peaks, preferably to baseline, is desired in order to achieve maximum purification. The speed at which any component of a mixture travels down the column in elution mode depends on many factors. But for two substances to travel at different speeds, and thereby be resolved, there must be substantial differences in some interaction between the biomolecules and the chromatography matrix. Operating parameters are adjusted to maximize the effect of this difference. In many cases, baseline separation of the peaks can be achieved only with gradient elution and low column loadings. Thus, two drawbacks to elution mode chromatography, especially at the preparative scale, are operational complexity, due to gradient solvent pumping, and low throughput, due to low column loadings. Displacement chromatography has advantages over elution chromatography in that components are resolved into consecutive zones of pure substances rather than “peaks”. Because the process takes advantage of the nonlinearity of the isotherms, a larger column feed can be separated on a given column with the purified components recovered at significantly higher concentration.

Reversed-phase Chromatography

Reversed phase HPLC (RP-HPLC) has a non-polar stationary phase and an aqueous, moderately polar mobile phase. One common stationary phase is a silica which has been surface-modified with RMe_2SiCl, where R is a straight chain alkyl group such as $C_{18}H_{37}$ or C_8H_{17}. With such stationary phases, retention time is longer for molecules which are less polar, while polar molecules elute more readily (early in the analysis). An investigator can increase retention times by adding more water to the mobile phase; thereby making the affinity of the hydrophobic analyte for the hydrophobic stationary phase stronger relative to the now more hydrophilic mobile phase. Similarly, an investigator can decrease retention time by adding more organic solvent to the eluent. RP-HPLC is so commonly used that it is often incorrectly referred to as "HPLC" without further specification. The pharmaceutical industry regularly employs RP-HPLC to qualify drugs before their release.

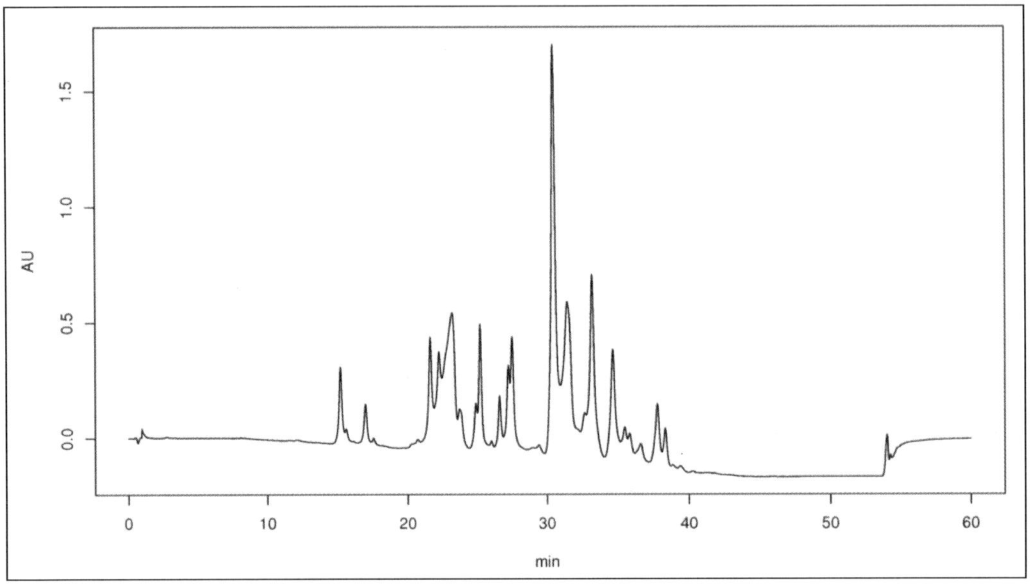

A chromatogram of complex mixture (perfume water) obtained by reversed phase HPLC

RP-HPLC operates on the principle of hydrophobic interactions, which originates from the high symmetry in the dipolar water structure and plays the most important role in all processes in life science. RP-HPLC allows the measurement of these interactive forces. The binding of the analyte to the stationary phase is proportional to the contact surface area around the non-polar segment of the analyte molecule upon association with the ligand on the stationary phase. This solvophobic effect is dominated by the force of water for "cavity-reduction" around the analyte and the C_{18}-chain versus the complex of both. The energy released in this process is proportional to the surface tension of the eluent (water: 7.3×10^{-6} J/cm^2, methanol: 2.2×10^{-6} J/cm^2) and to the hydrophobic surface of the analyte and the ligand respectively. The retention can be decreased by adding a less polar solvent (methanol, acetonitrile) into the mobile phase to reduce the surface tension of water. Gradient elution uses this effect by automatically reducing the polarity and the surface tension of the aqueous mobile phase during the course of the analysis.

Structural properties of the analyte molecule play an important role in its retention characteristics. In general, an analyte with a larger hydrophobic surface area (C–H, C–C, and generally non-polar atomic bonds, such as S-S and others) is retained longer because it is non-interacting with the water structure. On the other hand, analytes with higher polar surface area (conferred by the presence of polar groups, such as -OH, $-NH_2$, COO^- or $-NH_3^+$ in their structure) are less retained as they are better integrated into water. Such interactions are subject to steric effects in that very large molecules may have only restricted access to the pores of the stationary phase, where the interactions with surface ligands (alkyl chains) take place. Such surface hindrance typically results in less retention.

Retention time increases with hydrophobic (non-polar) surface area. Branched chain compounds elute more rapidly than their corresponding linear isomers because the overall surface area is decreased. Similarly organic compounds with single C–C bonds elute later than those with a C=C or C–C triple bond, as the double or triple bond is shorter than a single C–C bond.

Aside from mobile phase surface tension (organizational strength in eluent structure), other mobile phase modifiers can affect analyte retention. For example, the addition of inorganic salts causes a moderate linear increase in the surface tension of aqueous solutions (ca. 1.5×10^{-7} J/cm^2 per Mol for NaCl, 2.5×10^{-7} J/cm^2 per Mol for $(NH_4)_2SO_4$), and because the entropy of the analyte-solvent interface is controlled by surface tension, the addition of salts tend to increase the retention time. This technique is used for mild separation and recovery of proteins and protection of their biological activity in protein analysis (hydrophobic interaction chromatography, HIC).

Another important factor is the mobile phase pH since it can change the hydrophobic character of the analyte. For this reason most methods use a buffering agent, such as sodium phosphate, to control the pH. Buffers serve multiple purposes: control of pH, neutralize the charge on the silica surface of the stationary phase and act as ion pairing agents to neutralize analyte charge. Ammonium formate is commonly added in mass spectrometry to improve detection of certain analytes by the formation of analyte-ammonium adducts. A volatile organic acid such as acetic acid, or most commonly formic acid, is often added to the mobile phase if mass spectrometry is used to analyze the column effluent. Trifluoroacetic acid is used infrequently in mass spectrometry applications due to its persistence in the detector and solvent delivery system, but can be effective in improving retention of analytes such as carboxylic acids in applications utilizing other detectors, as it is a fairly strong organic acid. The effects of acids and buffers vary by application but generally improve chromatographic resolution.

Reversed phase columns are quite difficult to damage compared with normal silica columns; however, many reversed phase columns consist of alkyl derivatized silica particles and should never be used with aqueous bases as these will destroy the underlying silica particle. They can be used with aqueous acid, but the column should not be exposed to the acid for too long, as it can corrode the metal parts of the HPLC equipment. RP-HPLC columns should be flushed with clean solvent after use to remove residual acids or buffers, and stored in an appropriate composition of solvent. The metal content of HPLC columns must be kept low if the best possible ability to separate substances is to be retained. A good test for the metal content of a column is to inject a sample which is a mixture of 2,2'- and 4,4'-bipyridine. Because the 2,2'-bipy can chelate the metal, the shape of the peak for the 2,2'-bipy will be distorted (tailed) when metal ions are present on the surface of the silica.

Size-exclusion Chromatography

Size-exclusion chromatography (SEC), also known as gel permeation chromatography or gel filtration chromatography, separates particles on the basis of molecular size (actually by a particle's

Stokes radius). It is generally a low resolution chromatography and thus it is often reserved for the final, "polishing" step of the purification. It is also useful for determining the tertiary structure and quaternary structure of purified proteins. SEC is used primarily for the analysis of large molecules such as proteins or polymers. SEC works by trapping these smaller molecules in the pores of a particle. The larger molecules simply pass by the pores as they are too large to enter the pores. Larger molecules therefore flow through the column quicker than smaller molecules, that is, the smaller the molecule, the longer the retention time.

This technique is widely used for the molecular weight determination of polysaccharides. SEC is the official technique for the molecular weight comparison of different commercially available low-molecular weight heparins.

Ion-exchange Chromatography

In ion-exchange chromatography (IC), retention is based on the attraction between solute ions and charged sites bound to the stationary phase. Solute ions of the same charge as the charged sites on the column are excluded from binding, while solute ions of the opposite charge of the charged sites of the column are retained on the column. Solute ions that are retained on the column can be eluted from the column by changing the solvent conditions (e.g., increasing the ion effect of the solvent system by increasing the salt concentration of the solution, increasing the column temperature, changing the pH of the solvent, etc.).

Types of ion exchangers include polystyrene resins, cellulose and dextran ion exchangers (gels), and controlled-pore glass or porous silica. Polystyrene resins allow cross linkage which increases the stability of the chain. Higher cross linkage reduces swerving, which increases the equilibration time and ultimately improves selectivity. Cellulose and dextran ion exchangers possess larger pore sizes and low charge densities making them suitable for protein separation.

In general, ion exchangers favor the binding of ions of higher charge and smaller radius.

An increase in counter ion (with respect to the functional groups in resins) concentration reduces the retention time. A decrease in pH reduces the retention time in cation exchange while an increase in pH reduces the retention time in anion exchange. By lowering the pH of the solvent in a cation exchange column, for instance, more hydrogen ions are available to compete for positions on the anionic stationary phase, thereby eluting weakly bound cations.

This form of chromatography is widely used in the following applications: water purification, preconcentration of trace components, ligand-exchange chromatography, ion-exchange chromatography of proteins, high-pH anion-exchange chromatography of carbohydrates and oligosaccharides, and others.

Bioaffinity Chromatography

This chromatographic process relies on the property of biologically active substances to form stable, specific, and reversible complexes. The formation of these complexes involves the participation of common molecular forces such as the Van der Waals interaction, electrostatic interaction, dipole-dipole interaction, hydrophobic interaction, and the hydrogen bond. An efficient, biospecific bond is formed by a simultaneous and concerted action of several of these forces in the complementary binding sites.

Aqueous Normal-phase Chromatography

Aqueous normal-phase chromatography (ANP) is a chromatographic technique which encompasses the mobile phase region between reversed-phase chromatography (RP) and organic normal phase chromatography (ONP). This technique is used to achieve unique selectivity for hydrophilic compounds, showing normal phase elution using reversed-phase solvents.

Isocratic and Gradient Elution

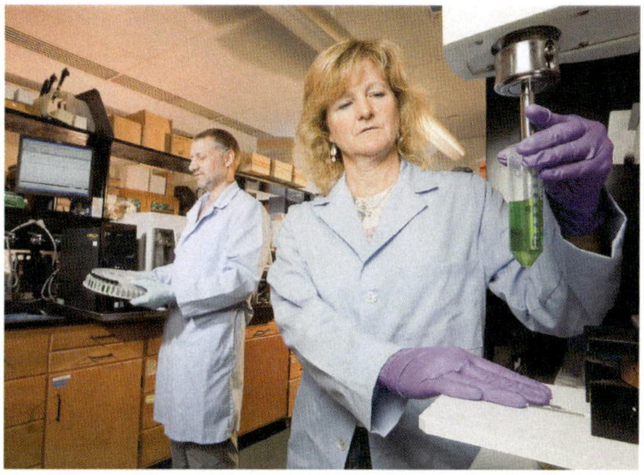

The support scientist (right) extracts plant pigments that will be analyzed by a plant physiologist (left) using an HPLC system.

A separation in which the mobile phase composition remains constant throughout the procedure is termed *isocratic* (meaning *constant composition*). The example of these the percentage of methanol throughout the procedure will remain constant i.e 10%. The word was coined by Csaba Horvath who was one of the pioneers of HPLC,

The mobile phase composition does not have to remain constant. A separation in which the mobile phase composition is changed during the separation process is described as a *gradient elution*. One example is a gradient starting at 10% methanol and ending at 90% methanol after 20 minutes. The two components of the mobile phase are typically termed "A" and "B"; *A* is the "weak" solvent which allows the solute to elute only slowly, while *B* is the "strong" solvent which rapidly elutes the solutes from the column. In reversed-phase chromatography, solvent *A* is often water or an aqueous buffer, while *B* is an organic solvent miscible with water, such as acetonitrile, methanol, THF, or isopropanol.

In isocratic elution, peak width increases with retention time linearly according to the equation for N, the number of theoretical plates. This leads to the disadvantage that late-eluting peaks get very flat and broad. Their shape and width may keep them from being recognized as peaks.

Gradient elution decreases the retention of the later-eluting components so that they elute faster, giving narrower (and taller) peaks for most components. This also improves the peak shape for tailed peaks, as the increasing concentration of the organic eluent pushes the tailing part of a peak forward. This also increases the peak height (the peak looks "sharper"), which is important in trace analysis. The gradient program may include sudden "step" increases in the percentage of the organic component, or different slopes at different times – all according to the desire for optimum separation in minimum time.

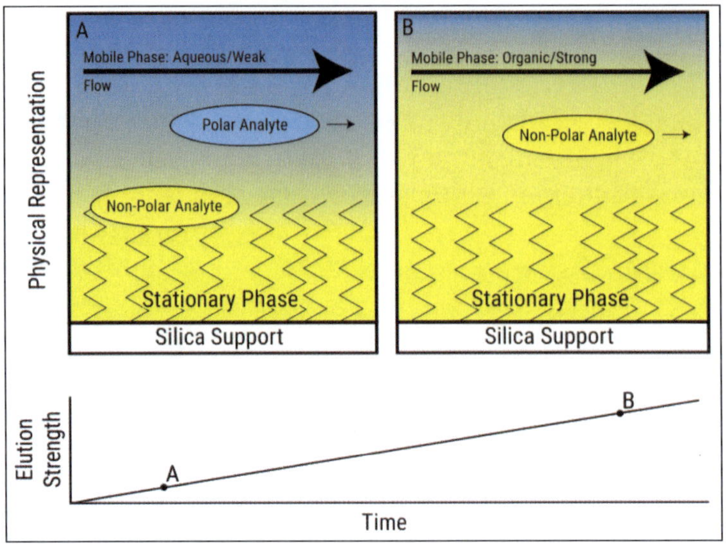

A schematic of gradient elution. Increasing mobile phase strength sequentially elutes analytes having varying interaction strength with the stationary phase.

In isocratic elution, the selectivity does not change if the column dimensions (length and inner diameter) change – that is, the peaks elute in the same order. In gradient elution, the elution order may change as the dimensions or flow rate change.

The driving force in reversed phase chromatography originates in the high order of the water structure. The role of the *organic component of the mobile phase* is to reduce this high order and thus *reduce the retarding strength of the aqueous component.*

Parameters

Theoretical

HPLC separations have theoretical parameters and equations to describe the separation of components into signal peaks when detected by instrumentation such as by a UV detector or a mass spectrometer. The parameters are largely derived from two sets of chromatagraphic theory: plate theory (as part of Partition chromatography), and the rate theory of chromatography/*Van Deemter equation*. Of course, they can be put in practice through analysis of HPLC chromatograms, although rate theory is considered the more accurate theory.

They are analogous to the calculation of retention factor for a paper chromatography separation, but describes how well HPLC separates a mixture into two or more components that are detected as peaks (bands) on a chromatogram. The HPLC parameters are the: efficiency factor(N), the retention factor (kappa prime), and the separation factor (alpha). Together the factors are variables in a resolution equation, which describes how well two components' peaks separated or overlapped each other. These parameters are mostly only used for describing HPLC reversed phase and HPLC normal phase separations, since those separations tend to be more subtle than other HPLC modes (e.g., ion exchange and size exclusion).

Void volume is the amount of space in a column that is occupied by solvent. It is the space within the column that is outside of the column's internal packing material. Void volume is measured on a chromatogram as the first component peak detected, which is usually the solvent that was present in the sample mixture; ideally the sample solvent flows through the column without interacting with the column, but is still detectable as distinct from the HPLC solvent. The void volume is used as a correction factor.

Efficiency factor (*N*) practically measures how sharp component peaks on the chromatogram are, as ratio of the component peak's area ("retention time") relative to the width of the peaks at their widest point (at the baseline). Peaks that are tall, sharp, and relatively narrow indicate that separation method efficiently removed a component from a mixture; high efficiency. Efficiency is very dependent upon the HPLC column and the HPLC method used. Efficiency factor is synonymous with plate number, and the 'number of theoretical plates'.

Retention factor (*kappa prime*) measures how long a component of the mixture stuck to the column, measured by the area under the curve of its peak in a chromatogram (since HPLC chromatograms are a function of time). Each chromatogram peak will have its own retention factor (e.g., *kappa1* for the retention factor of the first peak). This factor may be corrected for by the void volume of the column.

Separation factor (*alpha*) is a relative comparison on how well two neighboring components of the mixture were separated (*i.e.*, two neighboring bands on a chromatogram). This factor is defined in terms of a ratio of the retention factors of a pair of neighboring chromatogram peaks, and may also be corrected for by the void volume of the column. The greater the separation factor value is over 1.0, the better the separation, until about 2.0 beyond which an HPLC method is probably not needed for separation. Resolution equations relate the three factors such that high efficiency and separation factors improve the resolution of component peaks in a HPLC separation.

Internal Diameter

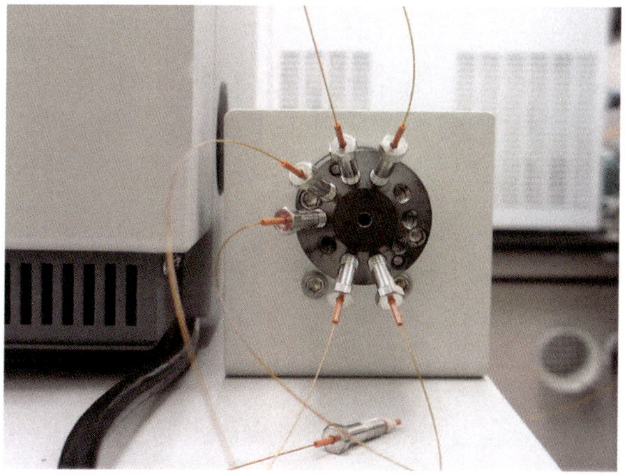

Tubing on a nano-liquid chromatography (nano-LC) system, used for very low flow capacities.

The internal diameter (ID) of an HPLC column is an important parameter that influences the detection sensitivity and separation selectivity in gradient elution. It also determines the quantity of analyte that can be loaded onto the column. Larger columns are usually seen in industrial applications, such as the purification of a drug product for later use. Low-ID columns have improved sensitivity and lower solvent consumption at the expense of loading capacity.

Larger ID columns (over 10 mm) are used to purify usable amounts of material because of their large loading capacity.

Analytical scale columns (4.6 mm) have been the most common type of columns, though smaller columns are rapidly gaining in popularity. They are used in traditional quantitative analysis of samples and often use a UV-Vis absorbance detector.

Narrow-bore columns (1–2 mm) are used for applications when more sensitivity is desired either with special UV-vis detectors, fluorescence detection or with other detection methods like liquid chromatography-mass spectrometry.

Capillary columns (under 0.3 mm) are used almost exclusively with alternative detection means such as mass spectrometry. They are usually made from fused silica capillaries, rather than the stainless steel tubing that larger columns employ.

Particle Size

Most traditional HPLC is performed with the stationary phase attached to the outside of small spherical silica particles (very small beads). These particles come in a variety of sizes with 5 µm beads being the most common. Smaller particles generally provide more surface area and better separations, but the pressure required for optimum linear velocity increases by the inverse of the particle diameter squared.

This means that changing to particles that are half as big, keeping the size of the column the same, will double the performance, but increase the required pressure by a factor of four. Larger particles are used in preparative HPLC (column diameters 5 cm up to >30 cm) and for non-HPLC applications such as solid-phase extraction.

Pore Size

Many stationary phases are porous to provide greater surface area. Small pores provide greater surface area while larger pore size has better kinetics, especially for larger analytes. For example, a protein which is only slightly smaller than a pore might enter the pore but does not easily leave once inside.

Pump Pressure

Pumps vary in pressure capacity, but their performance is measured on their ability to yield a consistent and reproducible volumetric flow rate. Pressure may reach as high as 60 MPa (6000 lbf/in2), or about 600 atmospheres. Modern HPLC systems have been improved to work at much higher pressures, and therefore are able to use much smaller particle sizes in the columns (<2 µm). These ultra high performance liquid chromatography" systems or UHPLCs can work at up to 120 MPa (17,405 lbf/in^2), or about 1200 atmospheres. The term "UPLC" is a trademark of the Waters Corporation, but is sometimes used to refer to the more general technique of UHPLC.

Detectors

HPLC detectors fall into two main categories: universal or selective. Universal detectors typically measure a bulk property (e.g., refractive index) by measuring a difference of a physical property between the mobile phase and mobile phase with solute while selective detectors measure a solute property (e.g., UV-Vis absorbance) by simply responding to the physical or chemical property of the solute. HPLC most commonly uses a UV-Vis absorbance detector, however, a wide range of other chromatography detectors can be used. A universal detector that complements UV-Vis absorbance detection is the Charged aerosol detector (CAD). A kind of commonly utilized detector includes refractive index detectors, which provide readings by measuring the changes in the refractive index of the effluent as it moves through the flow cell. In certain cases, it is possible to use multiple detectors, for example LCMS normally combines UV-Vis with a mass spectrometer.

Autosamplers

Large numbers of samples can be automatically injected onto an HPLC system, by the use of HPLC autosamplers. In addition, HPLC autosamplers have an injection volume and technique which is exactly the same for each injection, consequently they provide a high degree of injection volume precision.

Applications

Manufacturing

HPLC has many applications in both laboratory and clinical science. It is a common technique used in pharmaceutical development, as it is a dependable way to obtain and ensure product purity. While HPLC can produce extremely high quality (pure) products, it is not always the primary method used in the production of bulk drug materials. According to the European pharmacopoeia, HPLC is used in only 15.5% of syntheses. However, it plays a role in 44% of syntheses in the United States pharmacopoeia. This could possibly be due to differences in monetary and time constraints, as HPLC on a large scale can be an expensive technique. An increase in specificity, precision, and accuracy that occurs with HPLC unfortunately corresponds to an increase in cost.

Legal

This technique is also used for detection of illicit drugs in urine. The most common method of drug detection is an immunoassay. This method is much more convenient. However, convenience comes at the cost of specificity and coverage of a wide range of drugs. As HPLC is a method of determining (and possibly increasing) purity, using HPLC alone in evaluating concentrations of drugs is somewhat insufficient. With this, HPLC in this context is often performed in conjunction with mass spectrometry. Using liquid chromatography instead of gas chromatography in conjunction with MS circumvents the necessity for derivitizing with acetylating or alkylation agents, which can be a burdensome extra step. This technique has been used to detect a variety of agents like doping agents, drug metabolites, glucuronide conjugates, amphetamines, opioids, cocaine, BZDs, ketamine, LSD, cannabis, and pesticides. Performing HPLC in conjunction with Mass spectrometry reduces the absolute need for standardizing HPLC experimental runs.

Research

Similar assays can be performed for research purposes, detecting concentrations of potential clinical candidates like anti-fungal and asthma drugs. This technique is obviously useful in observing multiple species in collected samples, as well, but requires the use of standard solutions when information about species identity is sought out. It is used as a method to confirm results of synthesis reactions, as purity is essential in this type of research. However, mass spectrometry is still the more reliable way to identify species.

Medical

Medical use of HPLC can include drug analysis, but falls more closely under the category of nutrient analysis. While urine is the most common medium for analyzing drug concentrations, blood serum is the sample collected for most medical analyses with HPLC. Other methods of detection of molecules that are useful for clinical studies have been tested against HPLC, namely immunoassays. In one example of this, competitive protein binding assays (CPBA) and HPLC were compared for sensitivity in detection of vitamin D. Useful for diagnosing vitamin D deficiencies in children, it was found that sensitivity and specificity of this CPBA reached only 40% and 60%, respectively, of the capacity of HPLC. While an expensive tool, the accuracy of HPLC is nearly unparalleled.

Calorimetry

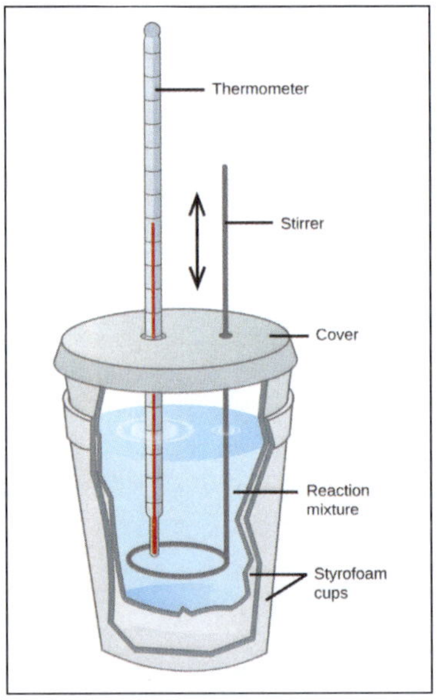

The world's first ice-calorimeter, used in the winter of 1782–83, by Antoine Lavoisier and Pierre-Simon Laplace, to determine the heat involved in various chemical changes; calculations which were based on Joseph Black's prior discovery of latent heat. These experiments mark the foundation of thermochemistry.

Calorimetry is the science or act of measuring changes in state variables of a body for the purpose of deriving the heat transfer associated with changes of its state due, for example, to chemical reactions, physical changes, or phase transitions under specified constraints. Calorimetry is performed with a calorimeter. Scottish physician and scientist Joseph Black, who was the first to recognize the distinction between heat and temperature, is said to be the founder of the science of calorimetry.

Snellen direct calorimetry chamber, University of Ottawa.

Indirect calorimetry calculates heat that living organisms produce by measuring either their production of carbon dioxide and nitrogen waste (frequently ammonia in aquatic organisms, or urea in terrestrial ones), or from their consumption of oxygen. Lavoisier noted in 1780 that heat production can be predicted from oxygen consumption this way, using multiple regression. The dynamic energy budget theory explains why this procedure is correct. Heat generated by living organisms may also be measured by direct calorimetry, in which the entire organism is placed inside the calorimeter for the measurement.

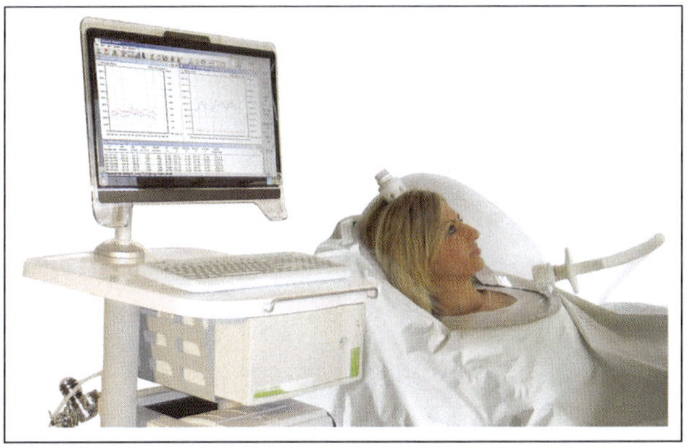

Indirect calorimetry metabolic cart measuring oxygen uptake and CO_2 production of a spontaneously breathing subject (dilution method with canopy hood).

A widely used modern instrument is the differential scanning calorimeter, a device which allows thermal data to be obtained on small amounts of material. It involves heating the sample at a controlled rate and recording the heat flow either into or from the specimen.

Classical Calorimetric Calculation of Heat

Cases with Differentiable Equation of State for a One-component Body

Calorimetry requires that a reference material that changes temperature have known definite thermal constitutive properties. The classical rule, recognized by Clausius and Kelvin, is that the pressure exerted by the calorimetric material is fully and rapidly determined solely by its temperature and volume; this rule is for changes that do not involve phase change, such as melting of ice. There are many materials that do not comply with this rule, and for them, the present formula of classical calorimetry does not provide an adequate account. Here the classical rule is assumed to hold for the calorimetric material being used, and the propositions are mathematically written:

> The thermal response of the calorimetric material is fully described by its pressure p as the value of its constitutive function $p(V,T)$ of just the volume V and the temperature T. All increments are here required to be very small. This calculation refers to a domain of volume and temperature of the body in which no phase change occurs, and there is only one phase present. An important assumption here is continuity of property relations. A different analysis is needed for phase change.

When a small increment of heat is gained by a calorimetric body, with small increments, δV of its volume, and δT of its temperature, the increment of heat, δQ , gained by the body of calorimetric material, is given by,

$$\delta Q = C_T^{(V)}(V,T)\delta V + C_V^{(T)}(V,T)\delta T$$

where,

$C_T^{(V)}(V,T)$ denotes the latent heat with respect to volume, of the calorimetric material at constant controlled temperature T.

The surroundings' pressure on the material is instrumentally adjusted to impose a chosen volume change, with initial volume V. To determine this latent heat, the volume change is effectively the independently instrumentally varied quantity. This latent heat is not one of the widely used ones, but is of theoretical or conceptual interest.

$C_V^{(T)}(V,T)$ denotes the heat capacity, of the calorimetric material at fixed constant volume V, while the pressure of the material is allowed to vary freely, with initial temperature T. The temperature is forced to change by exposure to a suitable heat bath. It is customary to write $C_V^{(T)}(V,T)$ simply as $C_V(V,T)$, or even more briefly as C_V. This latent heat is one of the two widely used ones.

The latent heat with respect to volume is the heat required for unit increment in volume at constant temperature. It can be said to be 'measured along an isotherm', and the pressure the material exerts is allowed to vary freely, according to its constitutive law $p = p(V,T)$. For a given material, it can have a positive or negative sign or exceptionally it can be zero, and this can depend on the temperature, as it does for water about 4 C. The concept of latent heat with respect to volume was perhaps first recognized by Joseph Black in 1762. The term 'latent heat of expansion' is also used. The latent heat with respect to volume can also be called the 'latent energy with respect to volume'. For all of these usages of 'latent heat', a more systematic terminology uses 'latent heat capacity'.

The heat capacity at constant volume is the heat required for unit increment in temperature at constant volume. It can be said to be 'measured along an isochor, and again, the pressure the material exerts is allowed to vary freely. It always has a positive sign. This means that for an increase in the temperature of a body without change of its volume, heat must be supplied to it. This is consistent with common experience.

Quantities like δQ are sometimes called 'curve differentials', because they are measured along curves in the (V,T) surface.

Classical Theory for Constant-volume: Isochoric Calorimetry

Constant-volume calorimetry is calorimetry performed at a constant volume. This involves the use of a constant-volume calorimeter. Heat is still measured by the above-stated principle of calorimetry.

This means that in a suitably constructed calorimeter, called a bomb calorimeter, the increment of volume δV can be made to vanish, $\delta V = 0$. For constant-volume calorimetry,

$$\delta Q = C_V \delta T$$

where,

δT denotes the increment in temperature,

C_V denotes the heat capacity at constant volume.

Classical Heat Calculation with Respect to Pressure

From the above rule of calculation of heat with respect to volume, there follows one with respect to pressure.

In a process of small increments, δp of its pressure, and δT of its temperature, the increment of heat, δQ, gained by the body of calorimetric material, is given by,

$$\delta Q = C_T^{(p)}(p,T)\delta p + C_p^{(T)}(p,T)\delta T$$

where,

$C_p^{(T)}(p,T)$ denotes the latent heat with respect to pressure, of the calorimetric material at constant temperature, while the volume and pressure of the body are allowed to vary freely, at pressure p and temperature T;

$C_T^{(p)}(p,T)$ denotes the heat capacity, of the calorimetric material at constant pressure, while the temperature and volume of the body are allowed to vary freely, at pressure p and temperature T. It is customary to write $C_p^{(T)}(p,T)$ simply as $C_p(p,T)$, or even more briefly as C_p.

The new quantities here are related to the previous ones,

$$C_T^{(p)}(p,T) = \frac{C_T^{(V)}(V,T)}{\left.\frac{\partial p}{\partial V}\right|_{(V,T)}}$$

$$C_p^{(T)}(p,T) = C_V^{(T)}(V,T) - C_T^{(V)}(V,T)\frac{\left.\frac{\partial p}{\partial T}\right|_{(V,T)}}{\left.\frac{\partial p}{\partial V}\right|_{(V,T)}}$$

where,

$\left.\frac{\partial p}{\partial V}\right|_{(V,T)}$ denotes the partial derivative of $p(V,T)$ with respect to V evaluated for (V,T),

and

$\left.\frac{\partial p}{\partial T}\right|_{(V,T)}$ denotes the partial derivative of $p(V,T)$ with respect to T evaluated for (V,T).

The latent heats $C_T^{(V)}(V,T)$ and $C_T^{(p)}(p,T)$ are always of opposite sign.

It is common to refer to the ratio of specific heats as,

$$\gamma(V,T) = \frac{C_p^{(T)}(p,T)}{C_V^{(T)}(V,T)} \text{ often just written as } \gamma = \frac{C_p}{C_V}.$$

Calorimetry through Phase Change, Equation of State Shows one Jump Discontinuity

An early calorimeter was that used by Laplace and Lavoisier, it worked at constant temperature, and at atmospheric pressure. The latent heat involved was then not a latent heat with respect to volume or with respect to pressure, as in the account for calorimetry without phase change. The latent heat involved in this calorimeter was with respect to phase change, naturally occurring at constant temperature. This kind of calorimeter worked by measurement of mass of water produced by the melting of ice, which is a phase change.

Cumulation of Heating

For a time-dependent process of heating of the calorimetric material, defined by a continuous joint progression $P(t_1,t_2)$ of $V(t)$ and $T(t)$, starting at time t_1 and ending at time t_2, there can be calculated an accumulated quantity of heat delivered, $\Delta Q(P(t_1,t_2))$. This calculation is done by mathematical integration along the progression with respect to time. This is because increments of heat are 'additive'; but this does not mean that heat is a conservative quantity. The idea that heat was a conservative quantity was invented by Lavoisier, and is called the 'caloric theory'; by the middle of the nineteenth century it was recognized as mistaken. Written with the symbol Δ, the quantity $\Delta Q(P(t_1,t_2))$ is not at all restricted to be an increment with very small values; this is in contrast with δQ.

One can write,

$$\Delta Q(P(t_1,t_2))$$

$$= \int_{P(t_1,t_2)} \dot{Q}(t)dt$$

$$= \int_{P(t_1,t_2)} C_T^{(V)}(V,T)\dot{V}(t)dt + \int_{P(t_1,t_2)} C_V^{(T)}(V,T)\dot{T}(t)dt.$$

This expression uses quantities such as $\dot{Q}(t)$.

Mathematical Aspects

The use of 'very small' quantities such as δQ is related to the physical requirement for the quantity $p(V,T)$ to be 'rapidly determined' by V and T; such 'rapid determination' refers to a physical process. These 'very small' quantities are used in the Leibniz approach to the infinitesimal calculus.

The Newton approach uses instead 'fluxions' such as $\dot{V}(t) = \left.\frac{dV}{dt}\right|_t$, which makes it more obvious that $p(V,T)$ must be 'rapidly determined'.

In terms of fluxions, the above first rule of calculation can be written,

$$\dot{Q}(t) = C_T^{(V)}(V,T)\dot{V}(t) + C_V^{(T)}(V,T)\dot{T}(t)$$

where,

t denotes the time,

$\dot{Q}(t)$ denotes the time rate of heating of the calorimetric material at time,

$\dot{V}(t)$ denotes the time rate of change of volume of the calorimetric material at time,

$\dot{T}(t)$ denotes the time rate of change of temperature of the calorimetric material.

The increment δQ and the fluxion $\dot{Q}(t)$ are obtained for a particular time t that determines the values of the quantities on the right hand sides of the above rules. But this is not a reason to expect that there should exist a mathematical function $Q(V,T)$. For this reason, the increment δQ is said to be an 'imperfect differential' or an 'inexact differential'. Some books indicate this by writing q instead of δQ. Also, the notation $đQ$ is used in some books. Carelessness about this can lead to error.

The quantity $\Delta Q(P(t_1,t_2))$ is properly said to be a functional of the continuous joint progression $P(t_1,t_2)$ of $V(t)$ and $T(t)$, but, in the mathematical definition of a function, $\Delta Q(P(t_1,t_2))$ is not a function of (V,T). Although the fluxion $\dot{Q}(t)$ is defined here as a function of time t, the symbols Q and $Q(V,T)$ respectively standing alone are not defined here.

Physical Scope of Calorimetry

These rules refer only to suitable calorimetric materials. The terms 'rapidly' and 'very small' call for empirical physical checking of the domain of validity of the above rules.

These rules for the calculation of heat belong to pure calorimetry. They make no reference to thermodynamics, and were mostly understood before the advent of thermodynamics. They are the basis of the 'thermo' contribution to thermodynamics. The 'dynamics' contribution is based on the idea of work.

Experimentally Conveniently Measured Coefficients

Empirically, it is convenient to measure properties of calorimetric materials under experimentally controlled conditions.

Pressure Increase at Constant Volume

For measurements at experimentally controlled volume, one can use the assumption, that the pressure of the body of calorimetric material is can be expressed as a function of its volume and temperature.

For measurement at constant experimentally controlled volume, the isochoric coefficient of pressure rise with temperature, is defined by:

$$\alpha_V(V,T) = \frac{\left.\frac{\partial p}{\partial V}\right|_{(V,T)}}{p(V,T)}.$$

Expansion at Constant Pressure

For measurements at experimentally controlled pressure, it is assumed that the volume V of the body of calorimetric material can be expressed as a function $V(T,p)$ of its temperature T and pressure p. This assumption is related to, but is not the same as, the above used assumption that the pressure of the body of calorimetric material is known as a function of its volume and temperature; anomalous behavior of materials can affect this relation.

The quantity that is conveniently measured at constant experimentally controlled pressure, the isobaric volume expansion coefficient, is defined by:

$$\beta_p(T,p) = \frac{\left.\frac{\partial V}{\partial T}\right|_{(T,p)}}{V(T,p)}.$$

Compressibility at Constant Temperature

For measurements at experimentally controlled temperature, it is again assumed that the volume V of the body of calorimetric material can be expressed as a function $V(T,p)$ of its temperature and pressure p, with the same provisos as mentioned just above.

The quantity that is conveniently measured at constant experimentally controlled temperature, the isothermal compressibility, is defined by

$$\kappa_T(T,p) = -\frac{\left.\frac{\partial V}{\partial p}\right|_{(T,p)}}{V(T,p)}.$$

Relation between Classical Calorimetric Quantities

Assuming that the rule $p = p(V,T)$ is known, one can derive the function of $\frac{\partial p}{\partial T}$ that is used in the classical heat calculation with respect to pressure. This function can be found experimentally from the coefficients $\beta_p(T,p)$ and $\kappa_T(T,p)$ through the mathematically deducible relation:

$$\frac{\partial p}{\partial T} = \frac{\beta_p(T,p)}{\kappa_T(T,p)}.$$

Connection between Calorimetry and Thermodynamics

Thermodynamics developed gradually over the first half of the nineteenth century, building on the theory of calorimetry which had been worked out before it, and on other discoveries. According to Gislason and Craig: “Most thermodynamic data come from calorimetry.” According to Kondepudi: “Calorimetry is widely used in present day laboratories.”

In terms of thermodynamics, the internal energy U of the calorimetric material can be considered as the value of a function $U(V,T)$ of (V,T), with partial derivatives $\frac{\partial U}{\partial V}$ and $\frac{\partial U}{\partial T}$.

Then it can be shown that one can write a thermodynamic version of the calorimetric rules,

$$\delta Q = \left[p(V,T) + \left.\frac{\partial U}{\partial V}\right|_{(V,T)} \right] \delta V + \left.\frac{\partial U}{\partial T}\right|_{(V,T)} \delta T$$

with,

$$C_T^{(V)}(V,T) = p(V,T) + \left.\frac{\partial U}{\partial V}\right|_{(V,T)}$$

and

$$C_V^{(T)}(V,T) = \left.\frac{\partial U}{\partial T}\right|_{(V,T)}$$

Again, further in terms of thermodynamics, the internal energy U of the calorimetric material can sometimes, depending on the calorimetric material, be considered as the value of a function $U(p,T)$ of (p,T), with partial derivatives $\frac{\partial U}{\partial p}$ and $\frac{\partial U}{\partial T}$, and with V being expressible as the value of a function $V(p,T)$ of (p,T), with partial derivatives $\frac{\partial V}{\partial p}$ and $\frac{\partial V}{\partial T}$.

Then, according to Adkins, it can be shown that one can write a further thermodynamic version of the calorimetric rules,

$$\delta Q = \left[\left.\frac{\partial U}{\partial p}\right|_{(p,T)} + p \left.\frac{\partial V}{\partial p}\right|_{(p,T)} \right] \delta p + \left[\left.\frac{\partial U}{\partial T}\right|_{(p,T)} + p \left.\frac{\partial V}{\partial T}\right|_{(p,T)} \right] \delta T$$

with,

$$C_T^{(p)}(p,T) = \left.\frac{\partial U}{\partial p}\right|_{(p,T)} + p\left.\frac{\partial V}{\partial p}\right|_{(p,T)}$$

and

$$C_p^{(T)}(p,T) = \left.\frac{\partial U}{\partial T}\right|_{(p,T)} + p\left.\frac{\partial V}{\partial T}\right|_{(p,T)}$$

Beyond the calorimetric fact the latent heats $C_T^{(V)}(V,T)$ and $C_T^{(p)}(p,T)$ are always of opposite sign, it may be shown, using the thermodynamic concept of work, that also,

$$C_T^{(V)}(V,T)\left.\frac{\partial p}{\partial T}\right|_{(V,T)} \geq 0.$$

Isothermal Segments of a Carnot Cycle

Calorimetry has a special benefit for thermodynamics. It tells about the heat absorbed or emitted in the isothermal segment of a Carnot cycle.

A Carnot cycle is a special kind of cyclic process affecting a body composed of material suitable for use in a heat engine. Such a material is of the kind considered in calorimetry, that exerts a pressure that is very rapidly determined just by temperature and volume. Such a body is said to change reversibly. A Carnot cycle consists of four successive stages or segments:

1. A change in volume from a volume V_a to a volume V_b at constant temperature T^+ so as to incur a flow of heat into the body (known as an isothermal change).
2. A change in volume from V_b to a volume V_c at a variable temperature just such as to incur no flow of heat (known as an adiabatic change).
3. Another isothermal change in volume from V_c to a volume V_d at constant temperature T^- such as to incur a flow or heat out of the body and just such as to precisely prepare for the following change.
4. Another adiabatic change of volume from V_d back to V_a just such as to return the body to its starting temperature T^+.

In isothermal segment (1), the heat that flows into the body is given by,

$$\Delta Q(V_a,V_b;T^+) = \int_{V_a}^{V_b} C_T^{(V)}(V,T^+)dV$$

and in isothermal segment (3) the heat that flows out of the body is given by,

$$-\Delta Q(V_c,V_d;T^-) =-\int_{V_c}^{V_d} C_T^{(V)}(V,T^-)dV$$

Because the segments (2) and (4) are adiabats, no heat flows into or out of the body during them, and consequently the net heat supplied to the body during the cycle is given by,

$$\Delta Q(V_a,V_b;T^+;V_c,V_d;T^-) = \Delta Q(V_a,V_b;T^+)+\Delta Q(V_c,V_d;T^-)$$
$$=\int_{V_a}^{V_b} C_T^{(V)}(V,T^+)dV+\int_{V_c}^{V_d} C_T^{(V)}(V,T^-)dV\,.$$

This quantity is used by thermodynamics and is related in a special way to the net work done by the body during the Carnot cycle. The net change of the body's internal energy during the Carnot cycle, $\Delta U(V_a,V_b;T^+;V_c,V_d;T^-)$, is equal to zero, because the material of the working body has the special properties.

Relations between Classical Calorimetric Quantities

Relation of Latent Heat with Respect to Volume and the Equation of State

The quantity $C_T^{(V)}(V,T)$, the latent heat with respect to volume, belongs to classical calorimetry. It accounts for the occurrence of energy transfer by work in a process in which heat is also transferred; the quantity, however, was considered before the relation between heat and work transfers was clarified by the invention of thermodynamics. In the light of thermodynamics, the classical calorimetric quantity is revealed as being tightly linked to the calorimetric material's equation of state $p=p(V,T)$. Provided that the temperature T is measured in the thermodynamic absolute scale, the relation is expressed in the formula:

$$C_T^{(V)}(V,T)=T\left.\frac{\partial p}{\partial T}\right|_{(V,T)}$$

Difference of Specific Heats

Advanced thermodynamics provides the relation:

$$C_p(p,T)-C_V(V,T)=\left[p(V,T)+\left.\frac{\partial U}{\partial V}\right|_{(V,T)}\right]\left.\frac{\partial V}{\partial T}\right|_{(p,T)}.$$

From this, further mathematical and thermodynamic reasoning leads to another relation between classical calorimetric quantities. The difference of specific heats is given by,

$$C_p(p,T)-C_V(V,T)=\frac{TV\beta_p^2(T,p)}{\kappa_T(T,p)}$$

Practical Constant-volume Calorimetry for Thermodynamic Studies

Constant-volume calorimetry is calorimetry performed at a constant volume. This involves the use of a constant-volume calorimeter.

No work is performed in constant-volume calorimetry, so the heat measured equals the change in internal energy of the system. The heat capacity at constant volume is assumed to be independent of temperature.

Heat is measured by the principle of calorimetry.

$$q = C_V \Delta T = \Delta U,$$

where,

ΔU is change in internal energy,

ΔT is change in temperature,

C_V is the heat capacity at constant volume.

In *constant-volume calorimetry* the pressure is not held constant. If there is a pressure difference between initial and final states, the heat measured needs adjustment to provide the *enthalpy change*. One then has,

$$\Delta H = \Delta U + \Delta(PV) = \Delta U + V\Delta P,$$

where,

ΔH is change in enthalpy,

V is the unchanging volume of the sample chamber.

Surface Analysis

The surface analysis is defined as the set of techniques which allow us to identify the atomic structure and chemistry state and composition of a material, the elements or compounds that are located on the surface of the material, and the interactions that take place in the interfaces of materials.

Using these surface analysis techniques, allows us to:

- Evaluate and optimize processes for surface preparation.
- Evaluate the system of protection against corrosion.
- Investigate and understand the processes of adhesion and rupture of the adhesives on the substrates.
- Analysis of failure in adhesive joints.
- Characterization and determination of the phenomena of ageing and degradation.

Among the surface analysis techniques can include:

- XPS - X-ray Photoelectron Spectroscopy,

- AES - Auger Electron Spectroscopy,
- Energy Dispersive X-EDX ray,
- SSIM - Spectroscopy of secondary ion mass,
- SPM - Scanning Probe Microscopy.

The XPS technique bases its operation on the emission of high-speed electrons which hit the surface and bounce with slower speed, an electron detector collects and analyzes the speed at which they bounced. Through this information we can identify and quantify components that form or that are in the surface of a material and its chemical state.

For example by using XPS we can identify and quantify the presence of silicon on a surface, which prevent good adhesion of the adhesive to the substrate.

The use of these techniques is very limited due to the high cost of purchasing the machines and the high degree of specialization needed for its operation and interpretation of data. But the information that gives us the use of these techniques can help to improve the quality and the productivity of adhesive processes.

In the future the idea would be to integrate these techniques during the production process, so that once the substrate surface preparation will assess their state for proceeding with the application of the adhesive, so that one obtains a quality process controlled and fully automated.

References

- Spectroscopy, science: britannica.com, Retrieved 8 January, 2019
- Sparkman, o. David (2000). Mass spectrometry desk reference. Pittsburgh: global view pub. Isbn 978-0-9660813-2-9
- Downard, kevin (2004). Mass spectrometry - a foundation course. Royal society of chemistry. Doi:10.1039/97 81847551306. Isbn 978-0-85404-609-6
- Schwartz jc, senko mw, syka je (june 2002). "a two-dimensional quadrupole ion trap mass spectrometer". Journal of the american society for mass spectrometry. 13 (6): 659–69. Doi:10.1016/s1044-0305(02)00384-7. Pmid 12056566
- Zolotov, yu. A. (2006). "ion mobility spectrometry". Journal of analytical chemistry. 61 (6): 519. Doi:10.1134/s1061934806060013
- Bard, allen j.; larry r. Faulkner (2000-12-18). Electrochemical methods: fundamentals and applications (2 ed.). Wiley. Isbn 978-0-471-04372-0
- Mcmurry, john (2011). Organic chemistry: with biological applications (2nd ed.). Belmont, ca: brooks/cole. P. 395. Isbn 9780495391470
- Surface-analysis: adhesiveandglue.com, Retrieved 13 May, 2019

Quantitative Analysis 3

- **Titration**
- **Gravimetric Analysis**
- **Kjeldahl Method**
- **Combustion Analysis**
- **Importance of Quantitative Analysis**

The branch of chemistry which is concerned with the determination of the percentage or amount of one or more constituents of a sample is referred to as quantitative analysis. A few of the methods of quantitative analysis include titration, gravimetric analysis, Kjeldahl method and combustion analysis. This chapter closely examines these different methods associated with quantitative analysis.

Quantitative chemical analysis is branch of chemistry that deals with the determination of the amount or percentage of one or more constituents of a sample. A variety of methods is employed for quantitative analyses, which for convenience may be broadly classified as chemical or physical, depending upon which properties are utilized. Chemical methods depend upon such reactions as precipitation, neutralization, oxidation, or, in general, the formation of a new compound. The major types of strictly chemical methods are known as gravimetric analysis and volumetric, or titrimetric, analysis. Physical methods involve the measurement of some physical property such as density, refractive index, absorption or polarization of light, electromotive force, magnetic susceptibility, and numerous others. An analysis will often require a combination of methods: qualitative for separating desired constituents from a sample and quantitative for measuring the amounts present.

The basic tool in all quantitative analyses is the analytical balance, used for the accurate weighing of samples and precipitates. For usual analytical work the balance should be able to determine differences in mass of 0.1 milligram (about 0.000004 ounce). In microanalyses the balance must be about 1,000 times more sensitive, and, for special work, balances of even higher sensitivity have been constructed.

Example: If you got an antacid powder sachet, you can notice the list of chemicals and also the quantity of each chemical present in the sachet. This quantity is approved by the government bodies and also the analysis of each constituent is carried as per the norms described in the respective nation pharmacopeia For instance if it is the United States, it will be done as per the United States Pharmacopeia (USP), in the UK by British Pharmacopeia (BP), etc.

"Qualitative analysis is meant to check/specify the actual chemical or substance and also chances of impurities if any." It relies on nature or chemistry of the compound. So the quality of the compound can be known.

Quantitative analysis indicates how much of the given substance or chemical is present in said sample. It relies on the concentration of the said compound in a given sample. Here the actual quantity of the compound in the sample is known.

In firms involved in the production of drugs, pharmaceuticals, nutraceuticals, chemicals, bio-molecules, etc. both qualitative and quantitative analysis are used to comply with the end product with desired or recommended standards.

Quantitative Analysis Methods

This is done by 3 basic methods by either.

1. Measuring the amount of reagent needed to complete the chemical reaction, or ascertaining the amount of reaction product obtained through suitable chemical reactions. Ex: Acid-base titrations.
2. Appropriate electrical measurements: Here the property of sample like its ability to conduct electricity or get oxidized or reduced are used as a function of quantity analysis.
3. Measurement of certain optical properties Some compounds have properties like the ability to absorb or emit a specific wavelength of light. This function of the compound is utilized for quantitative analysis of the substance.

Quantitative Analysis by Electrical Measurements

Amperometry here the sample which has the ability to conduct current is tested. Wherein the current and time needed to complete an electrochemical reaction is noted.

Potentiometry: Those substances which have an ability to oxidize or reduce are measured by this technique. In potentiometry, the changes in potential or EMF due to oxidation or reductive reaction are measured. During potentiometric titrations, a standard electrode and reference electrode is used. During the course of reaction progress, the standard electrode potential doesn't change but that of reference electrode changes. The addition of reacting reagent either oxidative (preferably) or reducing agent, changes the potential of the solution. The endpoint or equivalence point is indicated by a sudden change in potential of the graphical plot of electro-motive force (emf) plotted as a change in potential against the volume of the titrating solution.

Conductimetry which considers the electrical conductivity as a function of the quantity of substance of interest. Here the substance under test has the ability to conduct and on the addition of reacting substance, the conductivity goes down gradually till the endpoint. From there the conductance of externally added titrating agent rises.

Optical Methods of Quantitative Analysis

Absorption methods like visible spectroscopy i.e colorimetry, ultraviolet spectrophotometry, infrared spectrophotometry, etc. Here the substance in the sample absorbs a certain wavelength of light and this absorption is determined as a function of quantity.

Visible spectroscopy and UV spectroscopy rely on the light-absorbing property of the substance. Every substance absorbs light to a maximum extent at a specific wavelength. This wavelength of light is called lambda max.

Infrared spectroscopy is a bit different than the above. It is used to check the bonds in between the atoms and molecules.

Emission methods involve techniques like emission spectroscopy, fluorimetry, etc.. wherein the characteristic light emitted by a substance is recorded as a function of quantitative analysis.

Fluorimetry is based on the ability of a sample to absorb and re-emit light of a certain wavelength. When light is passed on to the sample at a specific wavelength, the electrons in the atoms get into exited state. They come back to the ground state by emitting light of a certain wavelength as fluorescence. This fluorescence emitted is measured to estimate the quantity of the sample.

Titration

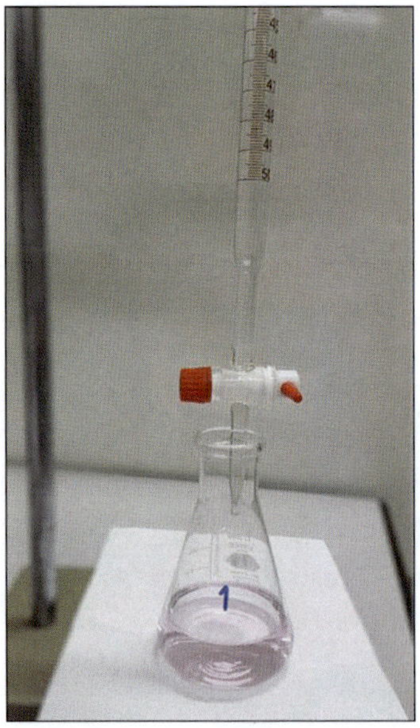

A burette and Erlenmeyer flask (conical flask) being used for an acid–base titration.

Titration (also called titrimetry and volumetric analysis) is a common laboratory method of quantitative chemical analysis to determine the concentration of an identified analyte (a substance to be analyzed). A reagent, called the titrant or titrator, is prepared as a standard solution of known concentration and volume. The titrant reacts with a solution of analyte (which may also be called the titrand) to determine the analyte's concentration. The volume of titrant that reacted with the analyte is called the titration volume.

Procedure

Analysis of soil samples by titration.

A typical titration begins with a beaker or Erlenmeyer flask containing a very precise volume of the analyte and a small amount of indicator (such as phenolphthalein) placed underneath a calibrated burette or chemistry pipetting syringe containing the titrant. Small volumes of the titrant are then added to the analyte and indicator until the indicator changes color in reaction to the titrant saturation threshold, reflecting arrival at the endpoint of the titration. Depending on the endpoint desired, single drops or less than a single drop of the titrant can make the difference between a permanent and temporary change in the indicator. When the endpoint of the reaction is reached, the volume of reactant consumed is measured and used to calculate the concentration of analyte by,

$$C_a = \frac{C_t V_t M}{V_a}$$

where C_a is the concentration of the analyte, typically in molarity; C_t is the concentration of the titrant, typically in molarity; V_t is the volume of the titrant used, typically in liters; M is the mole ratio of the analyte and reactant from the balanced chemical equation; and V_a is the volume of the analyte used, typically in liters.

Preparation Techniques

Typical titrations require titrant and analyte to be in a liquid (solution) form. Though solids are usually dissolved into an aqueous solution, other solvents such as glacial acetic acid or ethanol are used for special purposes (as in petrochemistry). Concentrated analytes are often diluted to improve accuracy.

Many non-acid–base titrations require a constant pH throughout the reaction. Therefore, a buffer solution may be added to the titration chamber to maintain the pH.

In instances where two reactants in a sample may react with the titrant and only one is the desired analyte, a separate masking solution may be added to the reaction chamber which masks the unwanted ion.

Some redox reactions may require heating the sample solution and titrating while the solution is still hot to increase the reaction rate. For instance, the oxidation of some oxalate solutions requires heating to 60 °C (140 °F) to maintain a reasonable rate of reaction.

Titration Curves

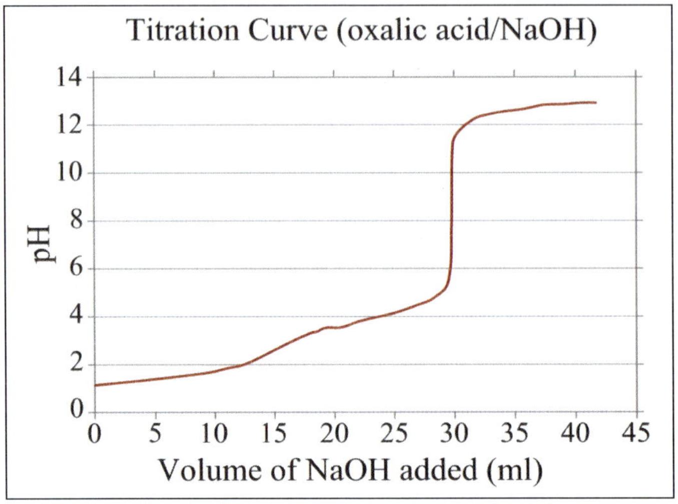

A typical titration curve of a diprotic acid titrated with a strong base. Shown here is oxalic acid titrated with sodium hydroxide. Both equivalence points are visible.

A titration curve is a curve in the plane whose *x*-coordinates are the volume of titrant added since the beginning of the titration, and whose *y*-coordinate is the concentration of the analyte at the corresponding stage of the titration (in an acid–base titration, the *y*-coordinate is usually the pH of the solution).

In an acid–base titration, the titration curve reflects the strength of the corresponding acid and base. For a strong acid and a strong base, the curve will be relatively smooth and very steep near the equivalence point. Because of this, a small change in titrant volume near the equivalence point results in a large pH change and many indicators would be appropriate (for instance litmus, phenolphthalein or bromothymol blue).

If one reagent is a weak acid or base and the other is a strong acid or base, the titration curve is irregular and the pH shifts less with small additions of titrant near the equivalence point. For example, the titration curve for the titration between oxalic acid (a weak acid) and sodium hydroxide (a strong base) is pictured. The equivalence point occurs between pH 8-10, indicating the solution is basic at the equivalence point and an indicator such as phenolphthalein would be appropriate. Titration curves corresponding to weak bases and strong acids are similarly behaved, with the solution being acidic at the equivalence point and indicators such as methyl orange and bromothymol blue being most appropriate.

Titrations between a weak acid and a weak base have titration curves which are highly irregular. Because of this, no definite indicator may be appropriate and a pH meter is often used to monitor the reaction.

The type of function that can be used to describe the curve is called a sigmoid function.

Types of Titrations

There are many types of titrations with different procedures and goals. The most common types of qualitative titration are acid–base titrations and redox titrations.

Acid–base Titration

Methyl orange.

Indicator	Color on acidic side	Range of color change (pH)	Color on basic side
Methyl violet	Yellow	0.0–1.6	Violet
Bromophenol blue	Yellow	3.0–4.6	Blue
Methyl orange	Red	3.1–4.4	Yellow
Methyl red	Red	4.4–6.3	Yellow
Litmus	Red	5.0–8.0	Blue
Bromothymol blue	Yellow	6.0–7.6	Blue
Phenolphthalein	Colorless	8.3–10.0	Pink
Alizarin yellow	Yellow	10.1–12.0	Red

Acid–base titrations depend on the neutralization between an acid and a base when mixed in solution. In addition to the sample, an appropriate pH indicator is added to the titration chamber, reflecting the pH range of the equivalence point. The acid–base indicator indicates the endpoint of the titration by changing color. The endpoint and the equivalence point are not exactly the same because the equivalence point is determined by the stoichiometry of the reaction while the endpoint is just the color change from the indicator. Thus, a careful selection of the indicator will reduce the indicator error. For example, if the equivalence point is at a pH of 8.4, then the Phenolphthalein indicator would be used instead of Alizarin Yellow because phenolphthalein would reduce the indicator error. Common indicators, their colors, and the pH range in which they change color are given in the table. When more precise results are required, or when the reagents are a weak acid and a weak base, a pH meter or a conductance meter are used.

For very strong bases, such as organolithium reagent, metal amides, and hydrides, water is generally not a suitable solvent and indicators whose pKa are in the range of aqueous pH changes are of little use. Instead, the titrant and indicator used are much weaker acids, and anhydrous solvents such as THF are used.

Phenolphthalein, a commonly used indicator in Acid and Base Titration.

The approximate pH during titration can be approximated by three kinds of calculations. Before beginning of titration, the concentration of $[H^+]$ is calculated in aqueous solution of weak acid before adding any base. When the number of moles of bases added equals the number of moles of initial acid or so called equivalence point, one of hydrolysis and the pH is calculated in the same way that the conjugate bases of the acid titrated was calculated. Between starting and end points, $[H^+]$ is obtained from the Henderson-Hasselbalch equation and titration mixture is considered as buffer. In Henderson-Hasselbalch equation the [acid] and [base] are said to be the molarities that would have been present even with dissociation or hydrolysis. In a buffer, $[H^+]$ can be exactly calculated but the dissociation of HA, the hydrolysis of A^- and self-ionization of water must be taken into account. Four independent equations must be used:

$$[H^+][OH^-] = 10^{-14}$$

$$[H^+] = K_a \frac{[HA]}{[A^-]}$$

$$[HA] + [A^-] = \frac{(n_A + n_B)}{V}$$

$$[H^+] + \frac{n_B}{V} = [A^-] + [OH^-]$$

In the equations, n_A and n_B are the moles of acid (HA) and salt (XA where X is the cation), respectively, used in the buffer, and the volume of solution is V. The law of mass action is applied to the ionization of water and the dissociation of acid to derived the first and second equations. The mass balance is used in the third equation, where the sum of $V[HA]$ and $V[A^-]$ must equal to the number of moles of dissolved acid and base, respectively. Charge balance is used in the fourth equation, where the left hand side represents the total charge of the cations and the right hand side represents the total charge of the anions: $\frac{n_B}{V}$ is the molarity of the cation (e.g. sodium, if sodium salt of the acid or sodium hydroxide is used in making the buffer).

Redox Titration

Redox titrations are based on a reduction-oxidation reaction between an oxidizing agent and a reducing agent. A potentiometer or a redox indicator is usually used to determine the endpoint of the titration, as when one of the constituents is the oxidizing agent potassium dichromate. The color change of the solution from orange to green is not definite, therefore an indicator such as sodium diphenylamine is used. Analysis of wines for sulfur dioxide requires iodine as an oxidizing agent. In this case, starch is used as an indicator; a blue starch-iodine complex is formed in the presence of excess iodine, signalling the endpoint.

Some redox titrations do not require an indicator, due to the intense color of the constituents. For instance, in permanganometry a slight persisting pink color signals the endpoint of the titration because of the color of the excess oxidizing agent potassium permanganate. In iodometry, at sufficiently large concentrations, the disappearance of the deep red-brown triiodide ion can itself be used as an endpoint, though at lower concentrations sensitivity is improved by adding starch indicator, which forms an intensely blue complex with triiodide.

Color of iodometric titration mixture before (left) and after (right) the end point.

Gas Phase Titration

Gas phase titrations are titrations done in the gas phase, specifically as methods for determining reactive species by reaction with an excess of some other gas, acting as the titrant. In one common gas phase titration, gaseous ozone is titrated with nitrogen oxide according to the reaction:

$$O_3 + NO \rightarrow O_2 + NO_2.$$

After the reaction is complete, the remaining titrant and product are quantified (e.g., by FT-IR); this is used to determine the amount of analyte in the original sample.

Gas phase titration has several advantages over simple spectrophotometry. First, the measurement does not depend on path length, because the same path length is used for the measurement of both the excess titrant and the product. Second, the measurement does not depend on a linear change in absorbance as a function of analyte concentration as defined by the Beer-Lambert law. Third, it is useful for samples containing species which interfere at wavelengths typically used for the analyte.

Complexometric Titration

Complexometric titrations rely on the formation of a complex between the analyte and the titrant. In general, they require specialized complexometric indicators that form weak complexes with the analyte. The most common example is the use of starch indicator to increase the sensitivity of iodometric titration, the dark blue complex of starch with iodine and iodide being more visible than iodine alone. Other complexometric indicators are Eriochrome Black T for the titration of calcium and magnesium ions, and the chelating agent EDTA used to titrate metal ions in solution.

Zeta Potential Titration

Zeta potential titrations are titrations in which the completion is monitored by the zeta potential, rather than by an indicator, in order to characterize heterogeneous systems, such as colloids. One of the uses is to determine the iso-electric point when surface charge becomes zero, achieved by changing the pH or adding surfactant. Another use is to determine the optimum dose for flocculation or stabilization.

Assay

An assay is a form of biological titration used to determine the concentration of a virus or bacterium. Serial dilutions are performed on a sample in a fixed ratio (such as 1:1, 1:2, 1:4, 1:8, etc.) until the last dilution does not give a positive test for the presence of the virus. The positive or negative value may be determined by visually inspecting the infected cells under a microscope or by an immunoenzymetric method such as enzyme-linked immunosorbent assay (ELISA). This value is known as the titer.

Measuring the Endpoint of a Titration

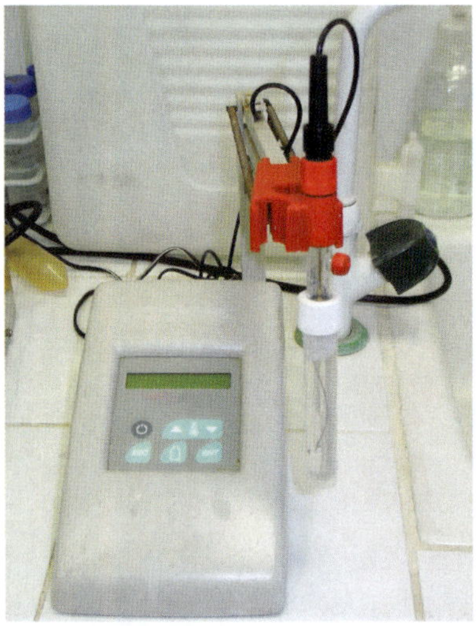

An elementary pH meter that can be used to monitor titration reactions.

Different methods to determine the endpoint include:

- Indicator: A substance that changes color in response to a chemical change. An acid–base indicator (e.g., phenolphthalein) changes color depending on the pH. Redox indicators are also used. A drop of indicator solution is added to the titration at the beginning; the endpoint has been reached when the color changes.
- Potentiometer: An instrument that measures the electrode potential of the solution. These are used for redox titrations; the potential of the working electrode will suddenly change as the endpoint is reached.

- pH meter: A potentiometer with an electrode whose potential depends on the amount of H^+ ion present in the solution. (This is an example of an ion-selective electrode.) The pH of the solution is measured throughout the titration, more accurately than with an indicator; at the endpoint there will be a sudden change in the measured pH.
- Conductivity: A measurement of ions in a solution. Ion concentration can change significantly in a titration, which changes the conductivity. (For instance, during an acid–base titration, the H^+ and OH^- ions react to form neutral H_2O.) As total conductance depends on all ions present in the solution and not all ions contribute equally (due to mobility and ionic strength), predicting the change in conductivity is more difficult than measuring it.
- Color change: In some reactions, the solution changes color without any added indicator. This is often seen in redox titrations when the different oxidation states of the product and reactant produce different colors.
- Precipitation: If a reaction produces a solid, a precipitate will form during the titration. A classic example is the reaction between Ag^+ and Cl^- to form the insoluble salt AgCl. Cloudy precipitates usually make it difficult to determine the endpoint precisely. To compensate, precipitation titrations often have to be done as "back" titrations.
- Isothermal titration calorimeter: An instrument that measures the heat produced or consumed by the reaction to determine the endpoint. Used in biochemical titrations, such as the determination of how substrates bind to enzymes.
- Thermometric titrimetry: Differentiated from calorimetric titrimetry because the heat of the reaction (as indicated by temperature rise or fall) is not used to determine the amount of analyte in the sample solution. Instead, the endpoint is determined by the rate of temperature change.
- Spectroscopy: Used to measure the absorption of light by the solution during titration if the spectrum of the reactant, titrant or product is known. The concentration of the material can be determined by Beer's Law.
- Amperometry: Measures the current produced by the titration reaction as a result of the oxidation or reduction of the analyte. The endpoint is detected as a change in the current. This method is most useful when the excess titrant can be reduced, as in the titration of halides with Ag^+.

Endpoint and Equivalence Point

Though equivalence point and endpoint are used interchangeably, they are different terms. *Equivalence point* is the theoretical completion of the reaction: the volume of added titrant at which the number of moles of titrant is equal to the number of moles of analyte, or some multiple thereof (as in polyprotic acids). *Endpoint* is what is actually measured, a physical change in the solution as determined by an indicator or an instrument.

There is a slight difference between the endpoint and the equivalence point of the titration. This error is referred to as an indicator error, and it is indeterminate.

Back Titration

Back titration is a titration done in reverse; instead of titrating the original sample, a known excess of standard reagent is added to the solution, and the excess is titrated. A back titration is useful if the endpoint of the reverse titration is easier to identify than the endpoint of the normal titration, as with precipitation reactions. Back titrations are also useful if the reaction between the analyte and the titrant is very slow, or when the analyte is in a non-soluble solid.

Graphical Methods

The titration process creates solutions with compositions ranging from pure acid to pure base. Identifying the pH associated with any stage in the titration process is relatively simple for monoprotic acids and bases. The presence of more than one acid or base group complicates these computations. Graphical methods, such as the equiligraph, have long been used to account for the interaction of coupled equilibria. These graphical solution methods are simple to implement, however they are infrequently used.

Particular Uses

A titration is demonstrated to secondary school students.

Acid–base Titrations

- In biodiesel: Waste vegetable oil (WVO) must be neutralized before a batch may be processed. A portion of WVO is titrated with a base to determine acidity, so the rest of the batch may be properly neutralized. This removes free fatty acids from the WVO that would normally react to make soap instead of biodiesel.
- Kjeldahl method: A measure of nitrogen content in a sample. Organic nitrogen is digested into ammonia with sulfuric acid and potassium sulfate. Finally, ammonia is back titrated with boric acid and then sodium carbonate.
- Acid value: The mass in milligrams of potassium hydroxide (KOH) required to titrate fully an acid in one gram of sample. An example is the determination of free fatty acid content.
- Saponification value: The mass in milligrams of KOH required to saponify a fatty acid in one gram of sample. Saponification is used to determine average chain length of fatty acids in fat.

- Ester value (or ester index): A calculated index. Ester value = Saponification value – Acid value.
- Amine value: The mass in milligrams of KOH equal to the amine content in one gram of sample.
- Hydroxyl value: The mass in milligrams of KOH corresponding to hydroxyl groups in one gram of sample. The analyte is acetylated using acetic anhydride then titrated with KOH.

Redox Titrations

- Winkler test for dissolved oxygen: Used to determine oxygen concentration in water. Oxygen in water samples is reduced using manganese(II) sulfate, which reacts with potassium iodide to produce iodine. The iodine is released in proportion to the oxygen in the sample, thus the oxygen concentration is determined with a redox titration of iodine with thiosulfate using a starch indicator.
- Vitamin C: Also known as ascorbic acid, vitamin C is a powerful reducing agent. Its concentration can easily be identified when titrated with the blue dye Dichlorophenolindophenol (DCPIP) which turns colorless when reduced by the vitamin.
- Benedict's reagent: Excess glucose in urine may indicate diabetes in the patient. Benedict's method is the conventional method to quantify glucose in urine using a prepared reagent. In this titration, glucose reduces cupric ions to cuprous ions which react with potassium thiocyanate to produce a white precipitate, indicating the endpoint.
- Bromine number: A measure of unsaturation in an analyte, expressed in milligrams of bromine absorbed by 100 grams of sample.
- Iodine number: A measure of unsaturation in an analyte, expressed in grams of iodine absorbed by 100 grams of sample.

Potentiometric Titration

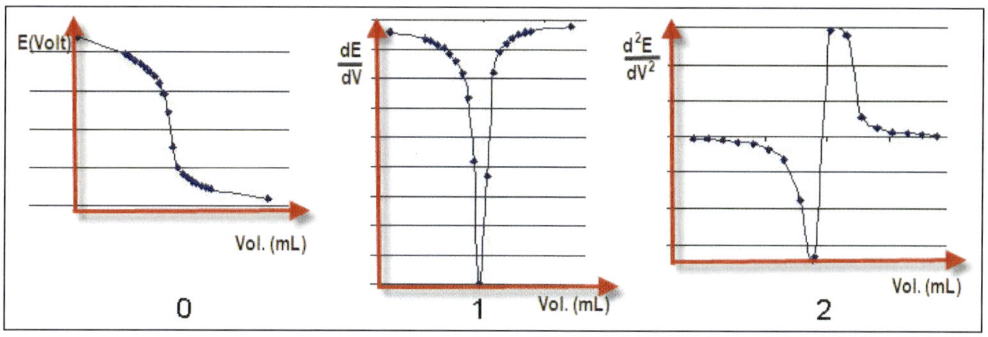

Measurements, first and second derivative in a potentiometric titration.

Potentiometric titration is a technique similar to direct titration of a redox reaction. It is a useful means of characterizing an acid. No indicator is used; instead the potential is measured across the analyte, typically an electrolyte solution. To do this, two electrodes are used, an indicator electrode (the glass electrode and metal ion indicator electrode) and a reference electrode. Reference electrodes generally used are hydrogen electrodes, calomel electrodes, and silver chloride electrodes.

The indicator electrode forms an electrochemical half cell with the interested ions in the test solution. The reference electrode forms the other half cell.

The overall electric potential is calculated as $E_{cell} = E_{ind} - E_{ref} + E_{sol}$. E_{sol} is the potential drop over the test solution between the two electrodes. E_{cell} is recorded at intervals as the titrant is added. A graph of potential against volume added can be drawn and the end point of the reaction is halfway between the jump in voltage. E_{cell} depends on the concentration of the interested ions with which the indicator electrode is in contact. For example, the electrode reaction may be:

$$M^{n+} + ne^{-} \rightarrow M.$$

As the concentration of M^{n+} changes, the E_{cell} changes correspondingly. Thus the potentiometric titration involve measurement of E_{cell} with the addition of titrant. Types of potentiometric titration: acid-base titration (total alkalinity and total acidity), redox titration (HI/HY and cerate), precipitation titration (halides), and complexometric titration.

Gravimetric Analysis

Gravimetric analysis describes a set of methods used in analytical chemistry for the quantitative determination of an analyte (the ion being analyzed) based on its mass. The principle behind this type of analysis is that once an ion's mass has been determined as a unique compound, that known measurement can then be used to determine the same analyte's mass in a mixture, as long as the relative quantities of the other constituents are known.

The four main types of this method of analysis are precipitation, volatilization, electro-analytical and miscellaneous physical method. The methods involve changing the phase of the analyte to separate it in its pure form from the original mixture and are quantitative measurements.

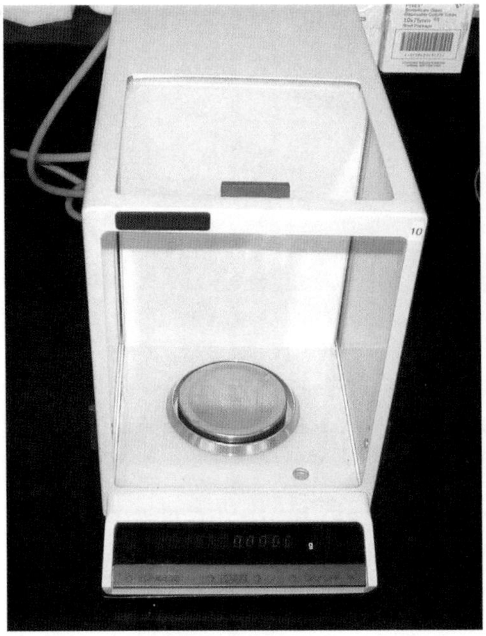

Analytical balance.

Precipitation Method

The precipitation method is the one used for the determination of the amount of calcium in water. Using this method, an excess of oxalic acid, $H_2C_2O_4$, is added to a measured, known volume of water. By adding a reagent, here ammonia, the calcium will precipitate as calcium oxalate. The proper reagent, when added to aqueous solution, will produce highly insoluble precipitates from the positive and negative ions that would otherwise be soluble with their counterparts.

The reaction is:

Formation of calcium oxalate:

$$Ca^{2+}(aq) + C_2O_4^{2-} \rightarrow CaC_2O_4$$

The precipitate is collected, dried and ignited to high (red) heat which converts it entirely to calcium oxide.

The reaction is pure calcium oxide formed,

$$CaC_2O_4 \rightarrow CaO(s) + CO(g) + CO_2(g)$$

The pure precipitate is cooled, then measured by weighing, and the difference in weights before and after reveals the mass of analyte lost, in this case calcium oxide. That number can then be used to calculate the amount, or the percent concentration, of it in the original mix.

Types of Volatilization Methods

In volatilization methods, removal of the analyte involves separation by heating or chemically decomposing a volatile sample at a suitable temperature. In other words, thermal or chemical energy is used to precipitate a volatile species. For example, the water content of a compound can be determined by vaporizing the water using thermal energy (heat). Heat can also be used, if oxygen is present, for combustion to isolate the suspect species and obtain the desired results.

The two most common gravimetric methods using volatilization are those for water and carbon dioxide. An example of this method is the isolation of sodium hydrogen bicarbonate (the main ingredient in most antacid tablets) from a mixture of carbonate and bicarbonate. The total amount of this analyte, in whatever form, is obtained by addition of an excess of dilute sulfuric acid to the analyte in solution.

In this reaction, nitrogen gas is introduced through a tube into the flask which contains the solution. As it passes through, it gently bubbles. The gas then exits, first passing a drying agent (here $CaSO_4$, the common desiccant *Drierite*). It then passes a mixture of the drying agent *and* sodium hydroxide which lays on asbestos or *Ascarite II*, a non-fibrous silicate containing sodium hydroxide. The mass of the carbon dioxide is obtained by measuring the increase in mass of this absorbent. This is performed by measuring the difference in weight of the tube in which the ascarite contained before and after the procedure.

The calcium sulfate ($CaSO_4$) in the tube retains carbon dioxide selectively as it's heated, and thereby, removed from the solution. The drying agent absorbs any aerosolized water and water vapor. The mix of the drying agent and NaOH absorbs the CO_2 and any water that may have been produced as a result of the absorption of the NaOH.

The reactions are:

Absorption of water:

$$NaHCO_3(aq) + H_2SO_4(aq) \rightarrow CO_2(g) + H_2O(l) + NaHSO_4(aq).$$

Absorption of CO_2 and residual water:

$$CO_2(g) + 2\ NaOH(s) \rightarrow Na_2CO_3(s) + H_2O(l).$$

Volatilization Methods

Volatilization methods can be either *direct* or *indirect*. Water eliminated in a quantitative manner from many inorganic substances by ignition is an example of a direct determination. It is collected on a solid desiccant and its mass determined by the gain in mass of the desiccant.

Another direct volatilization method involves carbonates which generally decompose to release carbon dioxide when acids are used. Because carbon dioxide is easily evolved when heat is applied, its mass is directly established by the measured increase in the mass of the absorbent solid used.

Determination of the amount of water by measuring the loss in mass of the sample during heating is an example of an indirect method. It is well known that changes in mass occur due to decomposition of many substances when heat is applied, regardless of the presence or absence of water. Because one must make the assumption that water was the only component lost, this method is less satisfactory than direct methods.

This often fault and misleading assumption has proven to be wrong on more than a few occasions. There are many substances other than water loss that can lead to loss of mass with the addition of heat, as well as a number of other factors that may contribute to it. The widened margin of error created by this all-too-often false assumption is not one to be lightly disregarded as the consequences could be far-reaching.

Nevertheless, the indirect method, although less reliable than direct, is still widely used in commerce. For example, it's used to measure the moisture content of cereals, where a number of imprecise and inaccurate instruments are available for this purpose.

Procedure

- The sample is dissolved, if it is not already in solution.
- The solution may be treated to adjust the pH (so that the proper precipitate is formed, or to suppress the formation of other precipitates). If it is known that species are present which interfere (by also forming precipitates under the same conditions as the analyte), the sample might require treatment with a different reagent to remove these interferents.
- The precipitating reagent is added at a concentration that favors the formation of a "good" precipitate. This may require low concentration, extensive heating often described as "digestion", or careful control of the pH. Digestion can help reduce the amount of coprecipitation.

- After the precipitate has formed and been allowed to "digest", the solution is carefully filtered. The filter is used to collect the precipitate; smaller particles are more difficult to filter.
    - Depending on the procedure followed, the filter might be a piece of ashless filter paper in a fluted funnel, or a filter crucible. Filter paper is convenient because it does not typically require cleaning before use; however, filter paper can be chemically attacked by some solutions (such as concentrated acid or base), and may tear during the filtration of large volumes of solution.
    - The alternative is a crucible whose bottom is made of some porous material, such as sintered glass, porcelain or sometimes metal. These are chemically inert and mechanically stable, even at elevated temperatures. However, they must be carefully cleaned to minimize contamination or carryover(cross-contamination). Crucibles are often used with a mat of glass or asbestos fibers to trap small particles.
    - After the solution has been filtered, it should be tested to make sure that the analyte has been completely precipitated. This is easily done by adding a few drops of the precipitating reagent; if a precipitate is observed, the precipitation is incomplete.
- After filtration, the precipitate – including the filter paper or crucible – is heated, or charred. This accomplishes the following:
    - The remaining moisture is removed (drying).
    - Secondly, the precipitate is converted to a more chemically stable form. For instance, calcium ion might be precipitated using oxalate ion, to produce calcium oxalate (CaC_2O_4); it might then be heated to convert it into the oxide (CaO). It is vital that the empirical formula of the weighed precipitate be known, and that the precipitate be pure; if two forms are present, the results will be inaccurate.
    - The precipitate cannot be weighed with the necessary accuracy in place on the filter paper; nor can the precipitate be completely removed from the filter paper to weigh it. The precipitate can be carefully heated in a crucible until the filter paper has burned away; this leaves only the precipitate. (As the name suggests, "ashless" paper is used so that the precipitate is not contaminated with ash.)
- After the precipitate is allowed to cool (preferably in a desiccator to keep it from absorbing moisture), it is weighed (in the crucible). To calculate the final mass of the analyte, the starting mass of the empty crucible is subtracted from the final mass of the crucible containing the sample. Since the composition of the precipitate is known, it is simple to calculate the mass of analyte in the original sample.

Example: A chunk of ore is to be analyzed for sulfur content. It is treated with concentrated nitric acid and potassium chlorate to convert all of the sulfur to sulfate (SO^{2-}_4). The nitrate and chlorate are removed by treating the solution with concentrated HCl. The sulfate is precipitated with barium (Ba^{2+}) and weighed as $BaSO_4$.

Advantages

Gravimetric analysis, if methods are followed carefully, provides for exceedingly precise analysis.

In fact, gravimetric analysis was used to determine the atomic masses of many elements in the periodic table to six figure accuracy. Gravimetry provides very little room for instrumental error and does not require a series of standards for calculation of an unknown. Also, methods often do not require expensive equipment. Gravimetric analysis, due to its high degree of accuracy, when performed correctly, can also be used to calibrate other instruments in lieu of reference standards.

Disadvantages

Gravimetric analysis usually only provides for the analysis of a single element, or a limited group of elements, at a time. Comparing modern dynamic flash combustion coupled with gas chromatography with traditional combustion analysis will show that the former is both faster and allows for simultaneous determination of multiple elements while traditional determination allowed only for the determination of carbon and hydrogen. Methods are often convoluted and a slight mis-step in a procedure can often mean disaster for the analysis (colloid formation in precipitation gravimetry, for example). Compare this with hardy methods such as spectrophotometry and one will find that analysis by these methods is much more efficient.

Steps in a Gravimetric Analysis

After appropriate dissolution of the sample the following steps should be followed for successful gravimetric procedure:

- Preparation of the Solution: This may involve several steps including adjustment of the pH of the solution in order for the precipitate to occur quantitatively and get a precipitate of desired properties, removing interferences, adjusting the volume of the sample to suit the amount of precipitating agent to be added.
- Precipitation: This requires addition of a precipitating agent solution to the sample solution. Upon addition of the first drops of the precipitating agent, supersaturation occurs, then nucleation starts to occur where every few molecules of precipitate aggregate together forming a nucleus. At this point, addition of extra precipitating agent will either form new nuclei or will build up on existing nuclei to give a precipitate. This can be predicted by Von Weimarn ratio where, according to this relation the particle size is inversely proportional to a quantity called the relative supersaturation where,

  Relative supersaturation = $(Q - S)/S$

  The Q is the concentration of reactants before precipitation, S is the solubility of precipitate in the medium from which it is being precipitated. Therefore, to get particle growth instead of further nucleation we must make the relative supersaturation ratio as small as possible. The optimum conditions for precipitation which make the supersaturation low are:

  - Precipitation using dilute solutions to decrease Q.
  - Slow addition of precipitating agent to keep Q as low as possible.
  - Stirring the solution during addition of precipitating agent to avoid concentration sites and keep Q low.

- Increase solubility by precipitation from hot solution.
- Adjust the pH to increase S, but not too much increase np as we do not want to lose precipitate by dissolution.
- Usually add a little excess of the precipitating agent for quantitative precipitation and check for completeness of the precipitation.

- Digestion of the precipitate: The precipitate is left hot (below boiling) for 30 min to one hour for the particles to be digested. Digestion involves dissolution of small particles and reprecipitation on larger ones resulting in particle growth and better precipitate characteristics. This process is called Ostwald ripening. An important advantage of digestion is observed for colloidal precipitates where large amounts of adsorbed ions cover the huge area of the precipitate. Digestion forces the small colloidal particles to agglomerate which decreases their surface area and thus adsorption. You should know that adsorption is a major problem in gravimetry in case of colloidal precipitate since a precipitate tends to adsorb its own ions present in excess, Therefore, forming what is called a primary ion layer which attracts ions from solution forming a secondary or counter ion layer. Individual particles repel each other keeping the colloidal properties of the precipitate. Particle coagulation can be forced by either digestion or addition of a high concentration of a diverse ions strong electrolytic solution in order to shield the charges on colloidal particles and force agglomeration. Usually, coagulated particles return to the colloidal state if washed with water, a process called peptization.
- Washing and Filtering the Precipitate: It is crucial to wash the precipitate thoroughly to remove all adsorbed species that would add to the weight of the precipitate. One should be careful nor to use too much water since part of the precipitate may be lost. Also, in case of colloidal precipitates we should not use water as a washing solution since peptization would occur. In such situations dilute nitric acid, ammonium nitrate, or dilute acetic acid may be used. Usually, it is a good practice to check for the presence of precipitating agent in the filtrate of the final washing solution. The presence of precipitating agent means that extra washing is required. Filtration should be done in appropriate sized Gooch or ignition filter paper.
- Drying and Ignition: The purpose of drying (heating at about 120-150 °C in an oven) or ignition in a muffle furnace at temperatures ranging from 600-1200 °C is to get a material with exactly known chemical structure so that the amount of analyte can be accurately determined.
- Precipitation from Homogeneous Solution: To make Q minimum we can, in some situations, generate the precipitating agent in the precipitation medium rather than adding it. For example, to precipitate iron as the hydroxide, we dissolve urea in the sample. Heating of the solution generates hydroxide ions from the hydrolysis of urea. Hydroxide ions are generated at all points in solution and thus there are no sites of concentration. We can also adjust the rate of urea hydrolysis and thus control the hydroxide generation rate. This type of procedure can be very advantageous in case of colloidal precipitates.

Solubility in the Presence of Diverse Ions

As expected from previous information, diverse ions have a screening effect on dissociated ions which leads to extra dissociation. Solubility will show a clear increase in presence of diverse ions as the solubility product will increase. Look at the following example:

Find the solubility of AgCl ($K_{sp} = 1.0 \times 10^{-10}$) in 0.1 M $NaNO_3$. The activity coefficients for silver and chloride are 0.75 and 0.76, respectively.

$$AgCl(s) = Ag^+ + Cl^-$$

We can no longer use the thermodynamic equilibrium constant (i.e. in absence of diverse ions) and we have to consider the concentration equilibrium constant or use activities instead of concentration if we use Kth:

$$K_{sp} = aAg^+ \, aCl^-$$

$$K_{sp} = [Ag^+] \, fAg^+ \, [Cl^-] \, fCl^-$$

$$1.0 \times 10^{-10} = s \times 0.75 \times s \times 0.76$$

$$s = 1.3 \times 10^{-5} \text{ M}$$

We have calculated the solubility of AgCl in pure water to be 1.0×10^{-5} M, if we compare this value to that obtained in presence of diverse ions we see % increase in solubility = $\{(1.3 \times 10^{-5} - 1.0 \times 10^{-5})/1.0 \times 10^{-5}\} \times 100 = 30\%$. Therefore, once again we have an evidence for an increase in dissociation or a shift of equilibrium to right in presence of diverse ions.

Kjeldahl Method

The Kjeldahl method or Kjeldahl digestion in analytical chemistry is a method for the quantitative determination of nitrogen contained in organic substances plus the nitrogen contained in the inorganic compounds ammonia and ammonium (NH_3/NH_4^+). Without modification, other forms of inorganic nitrogen, for instance nitrate, are not included in this measurement. This method was developed by Johan Kjeldahl in 1883.

Method

The method consists of heating a sample at 360–410°C with concentrated sulphuric acid (H_2SO_4), which decomposes ("digests") the organic sample by oxidation to liberate the reduced nitrogen as ammonium sulphate. Catalysts like selenium, Hg_2SO_4 or $CuSO_4$ are often added to hasten the digestion. Na_2SO_4 is also added to increase the boiling point of H_2SO_4. Digestion is complete when the liquor clarifies with the release of fumes. A distillation system depicted below is built.

The end of the condenser is dipped into a known volume of standard acid (i.e. acid of known concentration). A weak acid like boric acid (H_3BO_3) in excess of ammonia is often used. Standardized HCl, H_2SO_4 or some other strong acid can be used instead, but this is less commonplace. The sample solution is then distilled with a small amount of sodium hydroxide (NaOH). NaOH can also be added with a dropping funnel. NaOH reacts the ammonium (NH_4^+) to ammonia (NH_3), which boils off the sample solution. Ammonia bubbles through the standard acid solution and reacts back to ammonium salts with the weak or strong acid.

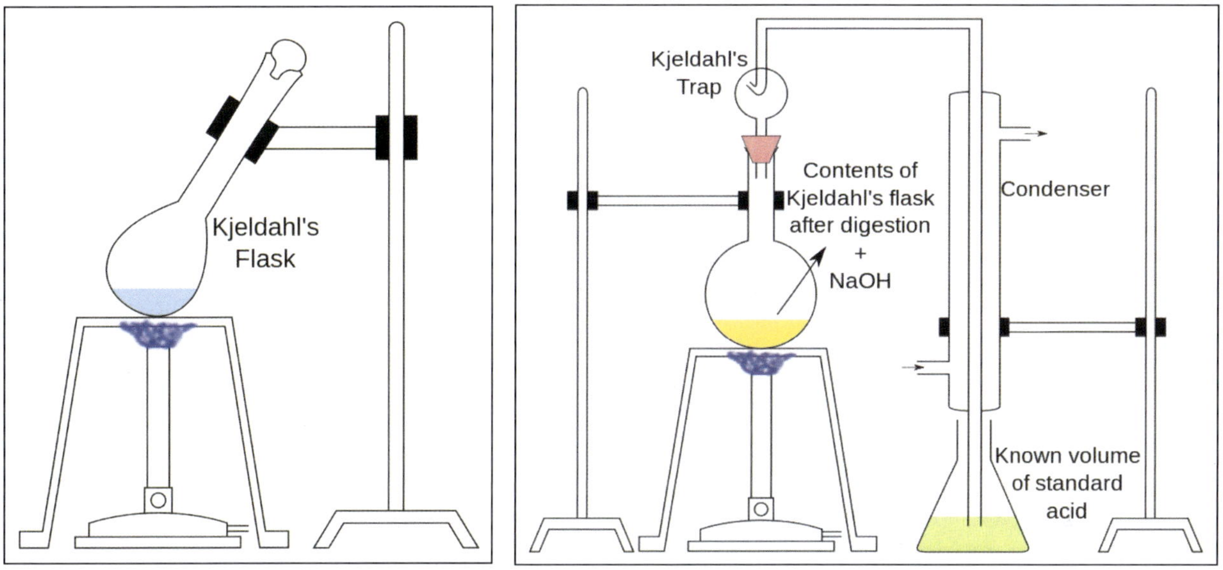

Ammonium ion concentration in the acid solution, and thus the amount of nitrogen in the sample, is measured via titration. If boric acid (or some other weak acid) was used, direct acid-base titration is done with a strong acid of known concentration. HCl or H_2SO_4 can be used. Indirect back titration is used instead if strong acids were used to make the standard acid solution: strong base of known concentration (like NaOH) is used to neutralize the solution. In this case, the amount of ammonia is calculated as the difference between the amount of HCl and NaOH. In the case of direct titration, it is not necessary to know the exact amount of weak acid (e.g. boric acid) because it does not interfere with the titration (it does have to be in excess of ammonia to efficiently trap it). Thus, one standard solution is needed (e.g. HCl) in the direct titration, while two are needed (e.g. HCl and NaOH) in the back-titration. One of the suitable indicators for these titration reactions is Tashiro's indicator.

In practice, this analysis is largely automated; specific catalysts accelerate the decomposition. Originally, the catalyst of choice was mercuric oxide. However, while it was very effective, health concerns resulted in it being replaced by cupric sulfate. Cupric sulfate was not as efficient as mercuric oxide, and yielded lower protein results. It was soon supplemented with titanium dioxide, which is currently the approved catalyst in all of the methods of analysis for protein in the Official Methods and Recommended Practices of AOAC International.

Applications

The Kjeldahl method's universality, precision and reproducibility have made it the internationally recognized method for estimating the protein content in foods and it is the standard method against which all other methods are judged. It is also used to assay soils, waste waters, fertilizers and other materials. It does not, however, give a measure of true protein content, as it measures nonprotein nitrogen in addition to the nitrogen in proteins. This is evidenced by the 2007 pet food incident and the 2008 Chinese milk powder scandal, when melamine, a nitrogen-rich chemical, was added to raw materials to fake high protein contents. Also, different correction factors are needed for different proteins to account for different amino acid sequences. Additional disadvantages, such as the need to use concentrated sulfuric acid at high temperature and the relatively long testing time (an hour or more), compare unfavorably with the Dumas method for measuring crude protein content.

Total Kjeldahl Nitrogen

Total Kjeldahl nitrogen or TKN is the sum of nitrogen bound in organic substances, nitrogen in ammonia (NH_3-N) and in ammonium (NH_4^+-N) in the chemical analysis of soil, water, or waste water (e.g. sewage treatment plant effluent).

Today, TKN is a required parameter for regulatory reporting at many treatment plants, and as a means of monitoring plant operations.

Conversion Factors

TKN is often used as a surrogate for protein in food samples. The conversion from TKN to protein depends on the type of protein present in the sample and what fraction of the protein is composed of nitrogenous amino acids, like arginine and histidine. However, the range of conversion factors is relatively narrow. Example conversion factors, known as N factors, for foods range from 6.38 for dairy and 6.25 for meat, eggs, maize (corn) and sorghum to 5.83 for most grains; 5.95 for rice, 5.70 for wheat flour, and 5.46 for peanuts. In practice, 6.25 is used for almost all food and feed regardless of applicability. The factor 6.25 is specifically required by US Nutrition Label regulations in the absence of another published factor.

Sensitivity

The Kjeldahl method is poorly sensitive in the original version. Other detection methods have been used to quantify NH_4^+ after mineralisation and distillation, achieving improved sensitivity: in-line generator of hydride coupled to a plasma atomic emission spectrometer (10–25 mg/L), potentiometric titration (>0.1 mg of nitrogen), zone capillary electrophoresis (1.5 µg/ml of nitrogen), and ion chromatography (0.5 µg/ml).

Limitations

Kjeldahl method is not applicable to compounds containing nitrogen in nitro and azo groups and nitrogen present in rings (e.g. pyridine, quinoline, isoquinoline) as nitrogen of these compounds does not convert to ammonium sulphate under the conditions of this method.

Combustion Analysis

Combustion analysis is a method used in both organic chemistry and analytical chemistry to determine the elemental composition (more precisely empirical formula) of a pure organic compound by combusting the sample under conditions where the resulting combustion products can be quantitatively analyzed. Once the number of moles of each combustion product has been determined the empirical formula or a partial empirical formula of the original compound can be calculated.

Applications for combustion analysis involve only the elements of carbon (C), hydrogen (H), nitrogen (N), and sulfur (S) as combustion of materials containing them convert these elements to their oxidized form (CO_2, H_2O, NO or NO_2, and SO_2) under high temperature high oxygen conditions. Notable interests for these elements involve measuring total nitrogen in food or feed to determine protein percentage, measuring sulfur in petroleum products, or measuring total organic carbon (TOC) in water.

Combustion Train

A combustion train is an analytical tool for the determination of elemental composition of a chemical compound. With knowledge of elemental composition a chemical formula can be derived. The combustion train allows the determination of carbon and hydrogen in a succession of steps:

- Combustion of the sample at high temperatures with Copper(II) oxide as the oxidizing agent,
- Collection of the resulting gas in a hygroscopic agent (magnesium perchlorate or calcium chloride) to trap generated water,
- Collection of the remainder gas in a strong base (for instance potassium hydroxide) to trap generated carbon dioxide.

Analytical determination of the amounts of water and carbon dioxide produced from a known amount of sample gives the empirical formula. For every hydrogen atom in the compound 1/2 equivalent of water is produced, and for every carbon atom in the compound 1 equivalent of carbon dioxide is produced.

Nowadays, modern instruments are sufficiently automated to be able to do these analyses routinely. Samples required are also extremely small — 0.5 mg of sample can be sufficient to give satisfactory CHN analysis.

Modern Methods

The water vapor, carbon dioxide and other products can be separated via gas chromatography and analysed via a thermal conductivity detector.

Importance of Quantitative Analysis

It's important to know the quantity of all or part of a sample for several reasons. If you're performing a chemical reaction, quantitative analysis helps you predict how much product to expect and to determine your actual yield.

Some reactions take place when the concentration of one component reaches a critical level. For example, an analysis of radioactive material might indicate there is enough of a key component for the specimen to undergo spontaneous fission.

Quantitative analysis is crucial to the formulation and testing of food and drugs, as it is used to measure nutrient levels and provide an accurate accounting of dosage.

It is also critical in determining the level of contaminants or the impurity of a sample. While qualitative analysis might be able to determine the presence of lead in the paint on a toy, for example, quantitative analysis detects how much concentration exists.

Medical tests rely on quantitative analysis for information about a patient's health. For example, quantitative analysis techniques can determine blood cholesterol levels or the ratio of lipoproteins in plasma or the amount of protein excreted in urine. Here again, quantitative analysis complements qualitative analysis, since the latter identifies the nature of a chemical while the former tells you how much there is.

Quantitative tests of a mineral may be used to determine whether it's practical to mine it for a specific element or compound. Quantitative tests are used to verify that products meet manufacturer or regulatory specifications.

References

- Quantitative-chemical-analysis, science: britannica.Com, retrieved 25 february, 2019
- Hannan, h.J. (2007). Technician's formulation handbook for industrial and household cleaning products. Lulu. Com. P. 103. Isbn 0-615-15601-0
- Quantitative-analysis-chemistry: studyread.Com, retrieved 16 january, 2019
- Harris, d.C. (2007). Quantitative chemical analysis (7 ed.). W. H. Freeman and company. Pp. 12. Isbn 978071 6770411
- Reger, d.L.; S.R. Goode; d.W. Ball (2009). Chemistry: principles and practice (3 ed.). Cengage learning. Pp. 684–693. Isbn 0-534-42012-5
- Yoder, claude (january 8, 2017). "Gravimetric analysis". Wiredchemist.Com. Retrieved january 8, 2017
- "Intérêt de l'ecz pour le dosage de l'azote total (méthode de kjeldahl) - blog pharma physic". Blog.Pharmaphysic.Fr. Retrieved 30 december 2017
- Definition-of-quantitative-analysis-604627: thoughtco.Com, retrieved 29 march, 2019

Qualitative Analysis

4

- **Flame Test**
- **Qualitative Analysis of Anions**
- **Qualitative Analysis of Cations**

The branch of chemistry that focuses on the identification or grouping of elements that are present in a sample is known as qualitative analysis. It can be broadly divided into quantitative analysis of anions and quantitative analysis of cations. The topics elaborated in this chapter will help in gaining a better perspective about these aspects of quantitative analysis.

Qualitative chemical analysis is branch of chemistry that deals with the identification of elements or grouping of elements present in a sample. The techniques employed in qualitative analysis vary in complexity, depending on the nature of the sample. In some cases it is necessary only to verify the presence of certain elements or groups for which specific tests applicable directly to the sample (e.g., flame tests, spot tests) may be available. More often the sample is a complex mixture, and a systematic analysis must be made in order that all the constituents may be identified. It is customary to classify the methods into two classes: qualitative inorganic analysis and qualitative organic analysis.

The classical procedure for the complete systematic analysis of an inorganic sample consists of several parts. First, a preliminary dry test may be performed, which may consist of heating the sample to detect the presence of such constituents as carbon (marked by the appearance of smoke or char) or water (marked by the appearance of moisture) or introducing the sample into a flame and noting the color produced. Certain elements may be identified by means of their characteristic flame colors. After preliminary tests have been performed, the sample is commonly dissolved in water for later determination of anionic constituents (*i.e.,* negatively charged elements or groupings of elements) and cationic constituents (*i.e.,* positively charged elements or groupings of elements). The procedure followed is based on the principle of treating the solution with a succession of reagents so that each reagent separates a group of constituents. The groups are then treated successively with reagents that divide a large group into subgroups or separate the constituents singly. When a constituent has been separated it is further examined to confirm its presence and to establish the amount present (quantitative analysis). Portions of the material are dissolved separately, and different procedures are used for each to detect the cationic and anionic constituents.

The organic nature of a compound is generally indicated by its behavior on being heated in air; solids usually melt, then burn with either a smoky or nonsmoky flame, in some instances leaving a black residue of carbon. The elements usually present in these compounds are carbon, hydrogen, oxygen, nitrogen, sulfur, and, occasionally, phosphorus, halogens, and some metals. Specific tests are available for each of the individual elements.

Lab Techniques for Semi-Micro Qualitative Analysis

It's easy to contaminate the sample through poor laboratory technique, so it's important to adhere to certain rules:

- Do not use tap water. Rather, use distilled water or deionized water.
- Glassware must be clean prior to use. It's not essential that it be dried.
- Don't put a reagent dropper tip into the mouth of a test tube. Dispense reagent from above the test tube lip to avoid contamination.
- Mix solutions by flicking the test tube. Never cover the test tube with a finger and shake the tube. Avoid exposing yourself to the sample.

Steps of Qualitative Analysis

- If the sample is presented as a solid (salt), it's important to note the shape and color of any crystals.
- Reagents are used to separate cations into groups of related elements.
- Ions in a group are separated from each other. After each separation stage, a test is performed to confirm certain ions truly were removed. The test is not performed on the original sample.
- Separations rely on different characteristics of ions. These may involve redox reactions to change oxidation state, differential solubility in an acid, base, or water, or precipitating certain ions.

Sample Qualitative Analysis Protocol

First, ions are removed in groups from the initial aqueous solution. After each group has been separated, then testing is conducted for the individual ions in each group. Here is a common grouping of cations:

- Group I: Ag^+, Hg_2^{2+}, Pb^{2+} Precipitated in 1 M HCl.
- Group II: Bi^{3+}, Cd^{2+}, Cu^{2+}, Hg^{2+}, (Pb^{2+}), Sb^{3+} and Sb^{5+}, Sn^{2+} and Sn^{4+} Precipitated in 0.1 M H_2S solution at pH 0.5.
- Group III: Al^{3+}, (Cd^{2+}), Co^{2+}, Cr^{3+}, Fe^{2+} and Fe^{3+}, Mn^{2+}, Ni^{2+}, Zn^{2+} Precipitated in 0.1 M H_2S solution at pH 9.
- Group IV: Ba^{2+}, Ca^{2+}, K^+, Mg^{2+}, Na^+, NH_4^+ Ba^{2+}, Ca^{2+}, and Mg^{2+} are precipitated in 0.2 M $(NH_4)_2CO_3$ solution at pH 10; the other ions are soluble.

Many reagents are used in the qualitative analysis, but only a few are involved in nearly every group procedure. The four most commonly used reagents are 6M HCl, 6M HNO_3, 6M NaOH, 6M NH_3. Understanding the uses of the reagents is helpful when planning an analysis.

Common Qualitative Analysis Reagents

Reagent	Effects
6M HCl	Increases $[H^+]$ Increases $[Cl^-]$ Decreases $[OH^-]$ Dissolves insoluble carbonates, chromates, hydroxides, some sulfates Destroys hydroxo and NH_3 complexes Precipitates insoluble chlorides
6M HNO_3	Increases $[H^+]$ Decreases $[OH^-]$ Dissolves insoluble carbonates, chromates, and hydroxides Dissolves insoluble sulfides by oxidizing sulfide ion Destroys hydroxo and ammonia complexes Good oxidizing agent when hot
6 M NaOH	Increases $[OH^-]$ Decreases $[H^+]$ Forms hydroxo complexes Precipitates insoluble hydroxides
6M NH_3	Increases $[NH_3]$ Increases $[OH^-]$ Decreases $[H^+]$ Precipitates insoluble hydroxides Forms NH_3 complexes Forms a basic buffer with NH_4^+

Flame Test

The flame test carried out on a copper halide. The characteristic bluish-green color of the flame is due to the copper.

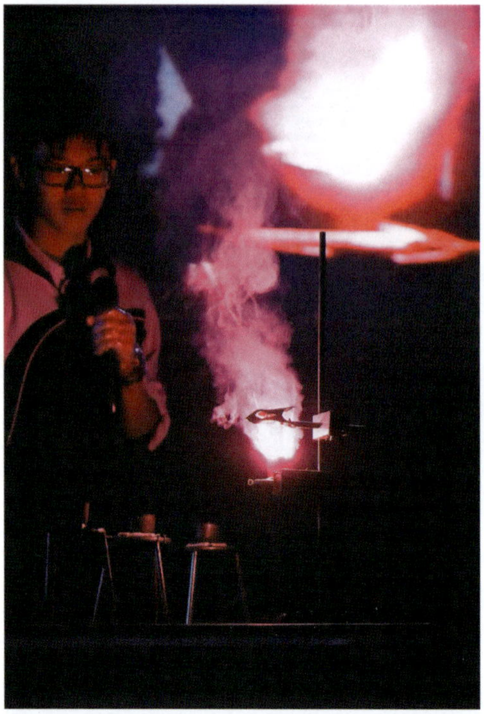

A flame test showing the presence of Lithium.

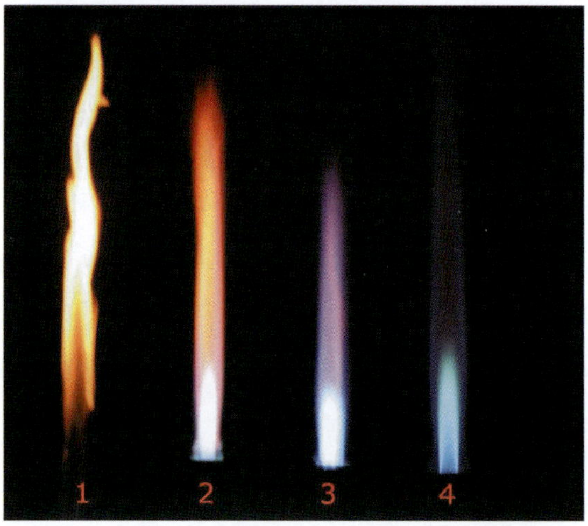

Different flame of Bunsen burner, depending on air flow through the valve:

1. Air valve closed.
2. Air valve nearly fully closed.
3. Air valve semi-opened.
4. Air valve maximally opened.

A *flame test* is an analytic procedure used in chemistry to detect the presence of certain elements, primarily metal ions, based on each element's characteristic emission spectrum. The color of flames in general also depends on temperature.

Gas flame.

Process

The test involves introducing a sample of the element or compound to a hot, non-luminous flame, and observing the color of the flame that results. The idea of the test is that sample atoms evaporate and since they are hot, they emit light when being in flame. Bulk sample emits light too, but its light is not good for analysis. Bulk sample emits light primarily due to the motion of the atoms, therefore its spectrum is broad, consisting of a broad range of colors. Separate atoms of a sample present in the flame can emit only due to electronic transitions between different atomic energy levels. Those transitions emit light of very specific frequencies, characteristic of the chemical element itself. Therefore, the flame gets the color, which is primarily determined by properties of the chemical element of the substance being put into flame. The flame test is a relatively easy experiment to set up and thus is often demonstrated or carried out in science classes in schools.

Samples are usually held on a platinum wire cleaned repeatedly with hydrochloric acid to remove traces of previous analytes. The compound is usually made into a paste with concentrated hydrochloric acid, as metal halides, being volatile, give better results. Different flames should be tried to avoid wrong data due to "contaminated" flames, or occasionally to verify the accuracy of the color. In high-school chemistry courses, wooden splints are sometimes used, mostly because solutions can be dried onto them, and they are inexpensive. Nichrome wire is also sometimes used. When using a splint, one must be careful to wave the splint through the flame rather than holding it in the flame for extended periods, to avoid setting the splint itself on fire. The use of cotton swab or melamine foam (used in "eraser" cleaning sponges) as a support have also been suggested.

Sodium is a common component or contaminant in many compounds and its spectrum tends to dominate over others. The test flame is often viewed through cobalt blue glass to filter out the yellow of sodium and allow for easier viewing of other metal ions.

Results

The flame test is relatively quick and simple to perform and can be carried out with the basic equipment found in most chemistry laboratories. However, the range of elements positively detectable under these conditions is small, as the test relies on the subjective experience of the experimenter rather than any objective measurements. The test has difficulty detecting small concentrations of some elements, while too strong a result may be produced for certain others, which tends to cause fainter colors to not appear.

Although the flame test only gives qualitative information, not quantitative data about the proportion of elements in the sample, quantitative data can be obtained by the related techniques of flame photometry or flame emission spectroscopy. Flame atomic absorption spectroscopy instruments, made by e.g. PerkinElmer or Shimadzu, can be operated in emission mode according to the instrument manuals.

Common Elements

Table: Common elements and their corresponding colors.

Symbol	Name	Color
Al	Aluminium	Silver-white, in very high temperatures such as an electric arc, light blue
As	Arsenic	Blue
B	Boron	Bright green
Ba	Barium	Pale/Apple green
Be	Beryllium	White
Bi	Bismuth	Azure
Ca	Calcium	Brick red, light green as seen through blue glass
Cd	Cadmium	Brick red
Ce	Cerium	Yellow
Co	Cobalt	Silver-white
Cr	Chromium	Silver-white
Cs	Caesium	Blue-violet
Cu(I)	Copper(I)	Bluish-green
Cu(II)	Copper(II) (non-halide)	Green
Cu(II)	Copper(II) (halide)	Blue-green
Ge	Germanium	Pale blue
Fe(II)	Iron(II)	Gold, when very hot such as an electric arc, bright blue, or green turning to orange-brown
Fe(III)	Iron(III)	Orange-brown
Hf	Hafnium	White
Hg	Mercury	Red
In	Indium	Indigo/Blue
K	Potassium	Lilac
Li	Lithium	crimson red; invisible through green glass
Mg	Magnesium	(none), but for burning Mg metal intense white
Mn (II)	Manganese (II)	Yellowish green
Mo	Molybdenum	Yellowish green
Na	Sodium	Intense yellow; invisible through cobalt blue glass

Nb	Niobium	Green or blue
Ni	Nickel	Silver-white (sometimes reported as colorless)
P	Phosphorus	Pale bluish green
Pb	Lead	Blue/white
Ra	Radium	Crimson
Rb	Rubidium	Red-violet
Sb	Antimony	Pale green
Sc	Scandium	Orange
Se	Selenium	Azure
Sn	Tin	Blue-white
Sr	Strontium	Crimson to Scarlet, yellowish through green glass and violet through blue cobalt glass
Ta	Tantalum	Blue
Te	Tellurium	Pale green
Ti	Titanium	Silver-white
Tl	Thallium	Pure green
V	Vanadium	Yellowish Green
W	Tungsten	Green
Y	Yttrium	Carmine, Crimson, or Scarlet
Zn	Zinc	Colorless (sometimes reported as bluish-green)
Zr	Zirconium	Mild red

Gold, silver, platinum, palladium, and a number of other elements do not produce a characteristic flame color, although some may produce sparks (as do metallic titanium and iron); salts of beryllium and gold reportedly deposit pure metal on cooling.

Qualitative Analysis of Anions

There is no systematic group classification for anions like metal cations. We can detect them by separate tests. For the ease of analysis they are divided into two broad classes. Class A and class B.

- Class A: Anions of class A can be identified by volatile products they evolve on treating with acids. This further can be divided into two subclasses.
    - Dilute acid group: This class of anions gives response with dilute hydrochloric or dilute sulphuric acid. These are Carbonate (CO_3^{2-}), Hydrogen carbonate or bicarbonate (HCO_3^-), Sulphite (SO_3^{2-}), Thiosulphate ($S_2O_3^{2-}$), Sulphide (S^{2-}), Nitrite (NO_2^-), Cyanide (CN^-), Cyanate (OCN^-), Hypochlorite (OCl^-).
    - Concentrated acid group: This class of anions gives response with concentrated sulphuric acid. These are Fluoride (F^-), Chloride (Cl^-), Bromide (Br^-), Iodide (I^-), Nitrate (NO_3^-), Sulphate (SO_4^{2-}), Acetate (CH_3COO^-), Oxalate ($(COO)_2^{2-}$), Chlorate (ClO_3^-), perchlorate

(ClO_4^-), Bromate (BrO_3^-), Thiocyanate (SCN^-), Citrate ($C_6H_5O_7^{3-}$), Tartrate ($C_4H_4O_6^{2-}$), permanganate (MnO_4^-), Hexacyanoferrate (I) and (II).

- Class B: Anions of class B are subdivided into two categories on the basis of reaction they give in solution.
  - Anions of this group give precipitation reaction in solution. Anions of this group are: Sulphate (SO_4^{2-}), Phosphate (PO_4^{3-}), Succinate ($C_4H_4O_4^{2-}$), Arsenate (AsO_4^{3-}), Chromate (CrO_4^{2-}), Silicate (SiO_3^{2-}), Salicylate ($C_6H_6(OH)COO^-$), Phosphite (HPO_3^{2-}), Arsenite (AsO_3^{3-}), Dichromate ($Cr_2O_7^{2-}$), Benzoate ($C_6H_6COO^-$).
  - Anions of this group give xxidation or reduction reaction in solution. Anions of this group are: Manganate (MnO_4^{2-}), Permanganate (MnO_4^-), Chromate (CrO_4^{2-}), Dichromate ($Cr_2O_7^{2-}$).

The scheme of analysis for anions is not as systematic as cations, that's why you may find some anions belong to more than one group. Before you start the analysis of anions, you have to separate them from their corresponding salt. There are two ways to separate anions from its metal cation partner. Either you can dissolve the salt into water or treat it with sodium carbonate solution.

Water Extract

Nitrites, nitrates, Thiosulphate and acetate are soluble in water. Water hydrolyses the salt and splits them into separate anions and cations. To prepare a water extract, take 1-2 g salt mixture in 10-20 ml of distilled water (because impurities may alter the dielectric constant of water and may affect its solvation ability) and boil for 2-3 minutes. Filter the residue and use filtrate as water extract for confirmatory tests of the anions.

Sodium Carbonate Extract

If the given salt mixture is partly soluble or insoluble in water then sodium carbonate extract is prepared for analysis. Preparation of soda extract is a three step process.

Step 1 - Dissolution: Take 1-2 g of salt mixture and 5-6 g anhydrous sodium carbonate, and add 10 ml water to dissolve them. Heat it for 15 minutes and filter. Wash the residue with water and add washings to the filtrate.

Step 2 - Acidification: Now acidify the filtrate with 8M nitric acid (HNO_3) and boil gently to expel CO_2(g) and allow it to cool.

Step 3 - Neutralization: Add 2M ammonia solution to make it just alkaline, stir the solution while adding ammonia. Now boil gently to remove excess of ammonia. Allow it to cool.

This is called as neutralized soda extract which we will use for the analysis of anions. On boiling with sodium carbonate, metal cations form metal carbonates that can be removed as precipitates and anions remain in solution in the company of sodium ions.

Since there is no systematic scheme for the analysis of anions, you have to analyze each of them separately. In the given mixture first try to find out which group of anion is present then try specific tests to confirm it. Take a small amount of given mixture in a test tube and add 2 ml dilute sulphuric acid and observe it, if there is no effervescence or bubbles then heat the solution and note the observation. Sign of any reaction with dilute sulphuric acid indicates the presence of dilute acid group anions. Preliminary observation can narrow down your search of anion, so observe very carefully.

Observation	Inference
Brisk effervescence and emission of colorless gas	Carbonate (CO_3^{2-}) or Hydrogen carbonate (HCO_3^-) may be present.
Colorless gas with rotten egg like smell	Sulphide (S^{2-}) may be present.
Colorless gas with a smell of burning sulphur	Sulphite (SO_3^{2-}) or Thiosulphate ($S_2O_3^{2-}$) may be present.
Colorless gas with vinegar like smell	Acetate (CH_3COO^-) may be present.
Colorless gas with odor of bitter almond	Cyanide (CN^-) may be present.
Light brown gas	Nitrite (NO_2^-) may be present.
Greenish colored gas with suffocating smell of chlorine	Hypochlorite (OCl^-) may be present.

Now try confirmatory test for each anion.

Carbonate (CO_3^{2-})

Carbonates are insoluble in water and we can use this information to detect them.

Lime water Test: Take a small amount of mixture and add dilute sulphuric acid, colorless gas will Be evolved with brisk effervescence (bubbles). If carbonate is present, it produces carbon dioxide gas.

$$H^+ + CO_3^{2-} \longrightarrow CO_2(g)\uparrow + H_2O$$

To identify CO_2 gas pass it to lime water, it produces turbidity in it due to formation of $CaCO_3$ which is insoluble in water.

$$CO_2(g) + Ca^{2+} + 2OH^- \longrightarrow CaCO_3\downarrow + H_2O$$

If you keep passing CO_2 in lime water turbidity may disappear due to formation of Hydrogen Carbonate which is soluble in water.

$$CaCO_3\downarrow + CO_2(g) + H_2O \longrightarrow Ca^{2+} + 2HCO_3^-$$

You might misinterpret carbonate with Hydrogen carbonate (bicarbonate) if you pass CO_2 for longer time, so you need another test which can distinguish between them. Similarly, Sulphur dioxide gas which is evolved due to presence of sulphite (SO_3^{2-}) also gives lime water test. It forms calcium sulphite $CaSO_3$. It's a white salt which is insoluble in water. If you want to test carbonate in presence of sulphite, you have to remove it first.

Lime Water Test for Carbonate in Presence of Sulphite

Take 0.5 g mixture in a test tube and add potassium dichromate solution to it, then add dilute sulphuric acid. Now collect the evolved gas and pass it into the lime water. Dichromate removes all SO_2 without affecting the CO_2.

$$SO_3^{2-} + H^+ \longrightarrow SO_2(g)\uparrow + H_2O$$

$$3SO_2 + Cr_2O_7^{2-} + 2H^+ \longrightarrow 2Cr^{3+} + 3SO_4^{2-} + H_2O$$

(Orange) (Green)

A redox reaction occurs between SO_2 and $Cr_2O_7^{2-}$, in which SO_2 oxidizes dichromate ion into Cr^{3+} and reduces itself to SO_4^{2-}. That's how you will be able to remove all SO_2 without affecting the CO_2.

How to distinguish between Carbonate (CO_3^{2-}) and Hydrogen carbonate or Bicarbonate (HCO_3^-)?

Take 0.5 g of mixture in a test tube and dissolve it in 4-5 ml of distilled water. Now divide the solution in two parts.

Part 1 - Test with Magnesium sulphate $MgSO_4$: By this test we can distinguish carbonate and bicarbonate on the basis of their reactivity. Carbonate is readily precipitated in cold solution while bicarbonate is precipitated only on boiling.

Add 1 ml $MgSO_4$ solution to the test solution. Carbonate readily forms magnesium carbonate.

$$CO_3^{2-} + Mg^{2+} \longrightarrow MgCO_3\downarrow$$

White precipitate confirms the presence of carbonate. If no precipitate is obtained in cold solution heat it to boiling. White precipitate is formed due to magnesium hydrogen carbonate. It confirms hydrogen carbonate.

$$HCO_3^- + Mg^{2+} \longrightarrow Mg(HCO_3)_2\downarrow$$

On further heating hydrogen carbonate dissociates and carbon dioxide gas is evolved. You can detect it with lime water test.

$$Mg(HCO_3)_2\downarrow \longrightarrow MgCO_3\downarrow + H_2O + CO_2\uparrow$$

Part 2 - Test with Calcium chloride $CaCl_2$: In this test, carbonate is easily precipitated while we need to add ammonia to confirm bicarbonate, which is precipitated as carbonate in the presence ammonia. Add excess of $CaCl_2$ solution to the test solution. Calcium carbonate is precipitated while calcium hydrogen carbonate remains in the solution. Filter it immediately as filtrate may contain bicarbonate HCO_3^-.

$$CO_3^{2-} + Ca^{2+} \longrightarrow CaCO_3\downarrow$$

Add ammonia to the filtrate. Ammonia is a base and HCO_3^- is a Brönsted acid. Ammonia NH_3 accepts hydrogen from hydrogen carbonate HCO_3^- and forms ammonium ion NH_4^+ (conjugate acid of NH_3) and carbonate CO_3^{2-} (conjugate base of HCO_3^-). This carbonate forms insoluble calcium carbonate which causes cloudiness into the solution. It confirms hydrogen carbonate.

$$HCO_3^- + Ca^{2+} + NH_3 \longrightarrow CaCO_3\downarrow + NH_4^+$$

Treatment with $AgNO_3$

The addition of Silver nitrate ($AgNO_3$) to your sample.

Some of the anions will form an insoluble solid (precipitate) with AgNO3 while other will remain soluble.

Followed by the addition of nitric acid (HNO_3) to the precipitate.

Some of the precipitates formed by the addition of $AgNO_3$ will dissolve when HNO_3 is added.

Followed by the addition of ammonia (NH_3), then acetic acid ($HC_2H_3O_2$), then silver nitrate ($AgNO_3$) to the supernatant.

Some of anions in solution will for form a precipitate again.

The results of these steps are summarized in the flow chart below.

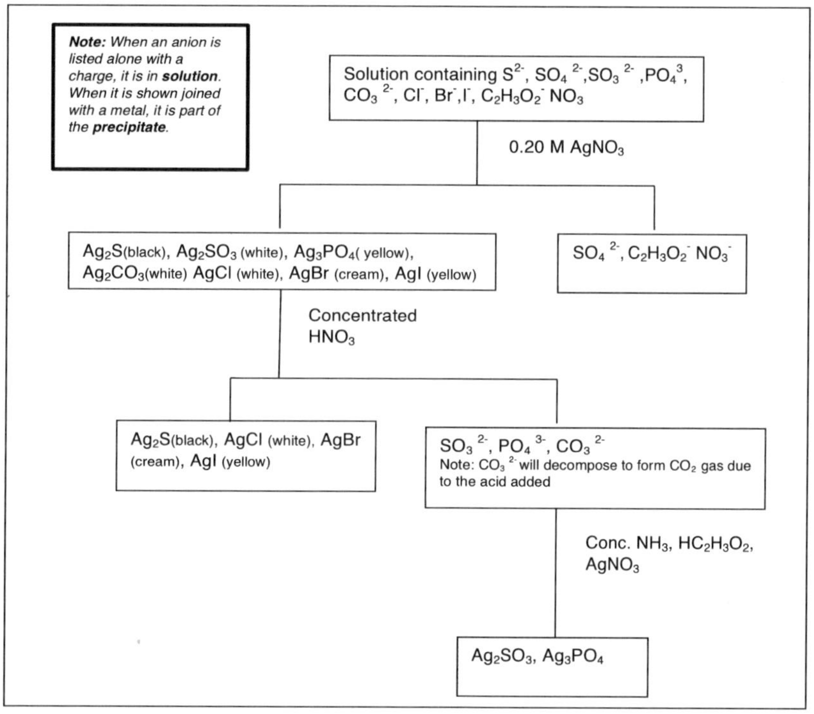

In order to understand why certain precipitates form, you may consult the solubility rules and Ksp values at the end of this experiment.

Treatment with $BaCl_2$

Starting with a new sample, the second set of preliminary tests involve the addition of barium chloride ($BaCl_2$). Again some of the anions will form precipitates while others will be soluble.

Addition of Barium Chloride ($BaCl_2$) to Sample

Some of the anions will form a precipitate with $BaCl_2$ while other will be soluble.

Followed by the addition of hydrochloric acid (HCl) to the precipitate.

Most of the precipitates formed by the addition of $BaCl_2$ will dissolve when HCl is added. $BaSO_4$ will remain as a solid.

The results of these steps are summarized in the flow chart below.

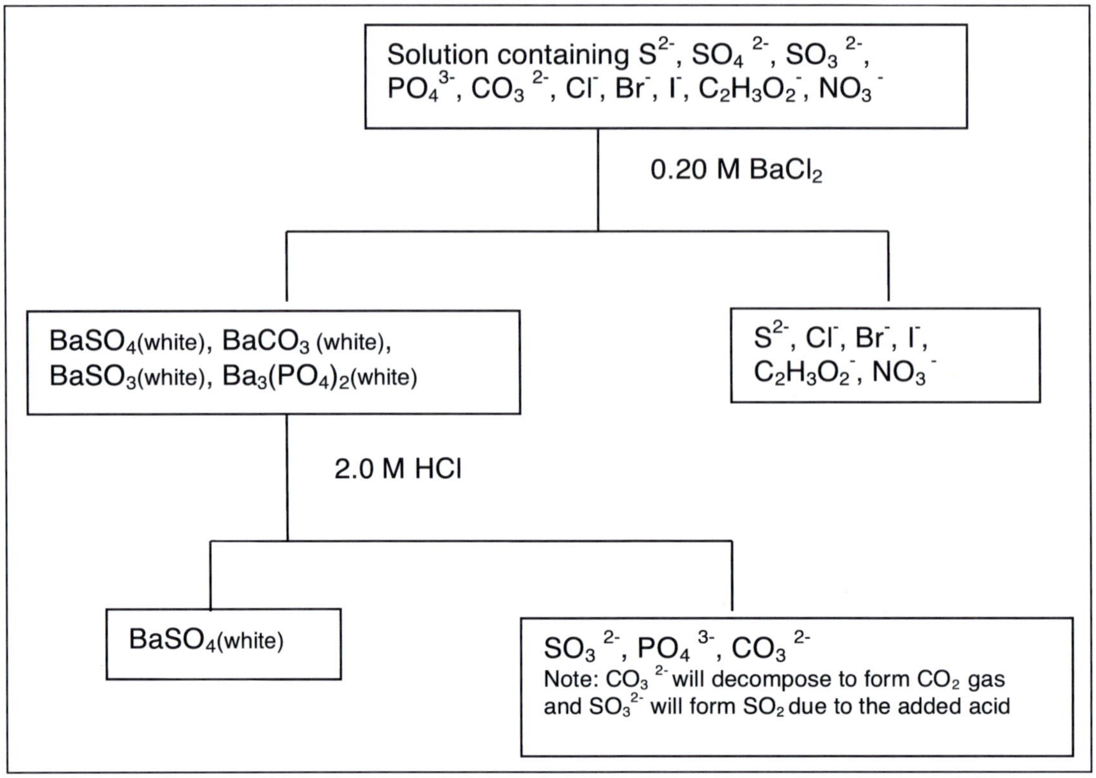

The addition of HCl in the second step fails to dissolve $BaSO_4$ since SO_4^{2-} is the anion of a strong acid.

Other anions also form insoluble precipitates with barium ions such as $BaCO_3$, $BaSO_3$, and $Ba_3(PO_4)_2$. However, theses are salts of weak acids and will all dissolve in acidic solution.

Salts of Weak Acids are Soluble in Strong Acids

To understand the different behavior of an anion from strong acid or a weak acid, recall Le Chatelier's Principle and consider a hypothetical salt MA which dissolves to form a cation M^+ and an anion A^-. Assume A is from a weak acid HA (more precisely the conjugate base of a weak acid HA). The fact that the acid is weak means that hydrogen ions (always present in aqueous solutions) and M^+ cations will both be competing for the A^-:

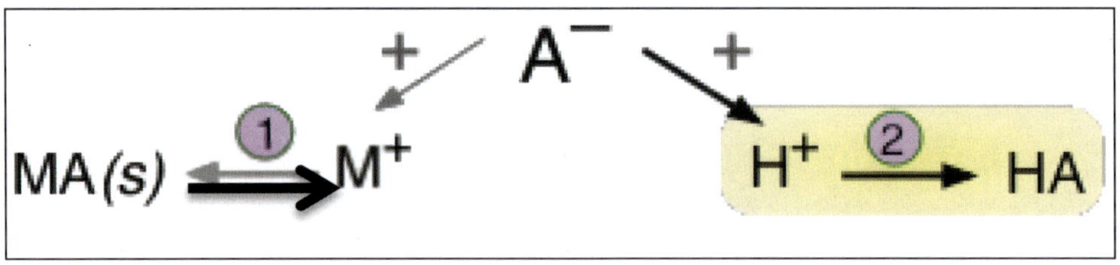

The weaker the acid HA, the more reaction 's equilibrium lies to the right. This reaction will gobble up A^- ions. If an excess of H^+ is made available by the addition of a strong acid, even more A^- ions will be consumed. This will cause reaction 's equilibrium to shift to the right and the solid to dissolve.

Strong Acids will not Dissolve Salts of Strong Acids

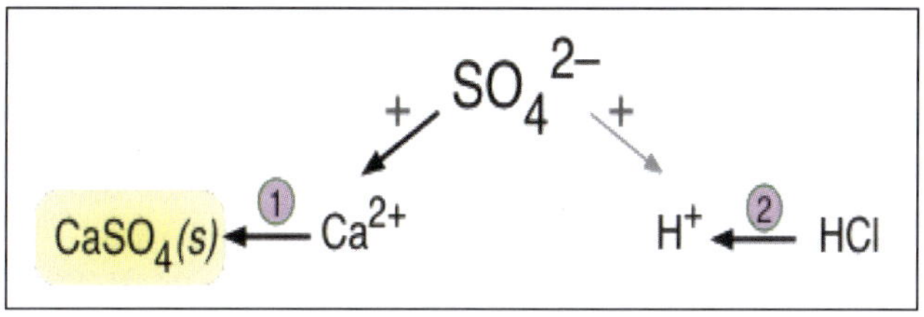

Consider the addition of a strong acid to a solution containing the salt $CaSO_4$. Recall that the anion of this salt (SO_4^{2-}) is part of the strong acid H_2SO_4.

In sulfate ions react with calcium ions to form insoluble $CaSO_4$. The addition of a strong acid such as HCl (which will totally dissociated to form H^+ ions) has no effect because H^+ and SO_4^{2-} will not combine into a the strong acid H_2SO_4.

Note: Although H^+ can protonate some SO_4^{2-} ions to form hydrogen sulfate ("bisulfate") HSO4 –, this ampholyte acid is too weak to reverse.

If the ions are from a weak acid/base, then changing the pH will change the solubility; if the ions are from a strong acid/base, then pH does not change solubility.

Example: Which solid, AgF(s), AgCl(s), AgBr(s), or AgI(s), is more soluble in an acidic solution (as compared to pure water)?

To understand, this write the reaction of H^+ with AgCl: $AgCl(s) + H^+(aq) \rightleftharpoons Ag^+(aq) + HCl(aq)$; since HCl can not form in water since it is a strong acid (strong acids dissociate 100% in water), this reaction does not proceed.

Repeat with AgF: $AgF(s) + H^+(aq) \rightleftharpoons Ag^+(aq) + HF(aq)$; since HF is a weak acid, it can form in water and the reaction can proceed and the AgF dissolves more in an acidic solution.

Treatment with H_2SO_4

Starting with a new sample, the third elimination tests involve the addition of 2.0 M sulfuric acid. Some of the anions will react to form a gas while others will undergo no reaction.

The results of these steps are summarized in the flow chart below.

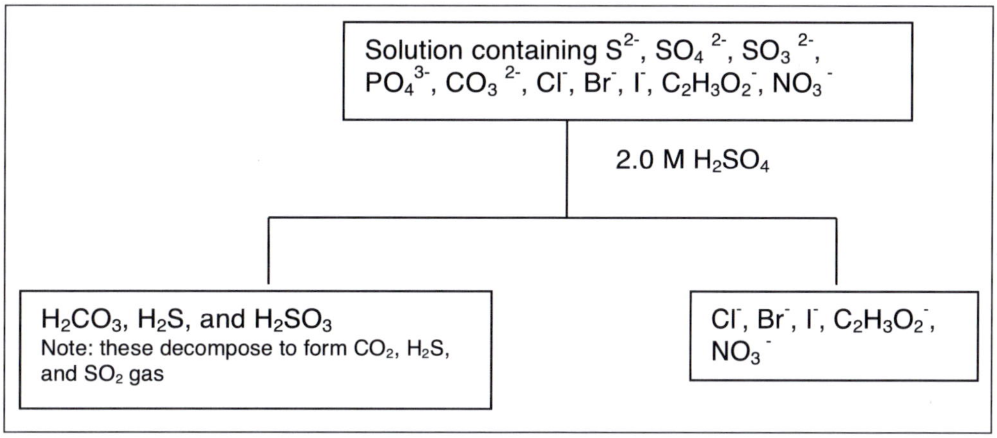

The anions CO_3^{-2}, S^{2-}, and SO_3^{2-} are the anions of the weak acids H_2CO_3, H_2S, and H_2SO_3 respectively. These weak acids are unstable in solution and decompose to produce the gases CO_2, H_2S, and SO_2 respectively.

$$SO_3^{2-} + 2\,H_3O^+ \rightarrow SO_2(g) + 3H_2O$$

$$S^{2-} + 2\,H_3O^+ \rightarrow H_2S(g) + 2H_2O$$

$$CO_3^{2-} + 2\,H_3O^+ \rightarrow CO_2(g) + 3H_2O$$

If no gases are produced from the acidic solution, then these three anions are absent.

Procedure

Safety

Most of the acids and bases used are very concentrated and can cause chemical burns if spilled. Handle them with care. Wash acid or base spills off of yourself with lots of water. Small spills (a few drops) can be cleaned up with paper towels. Larger acid spills can be neutralized with baking soda, $NaHCO_3$, and then safely cleaned up. Neutralize base spills with a vinegar solution (dilute acetic acid).

Solutions containing silver ions cause stains which do not appear immediately. If you suspect that you spilled any of these solutions on yourself, wash off with soap and water.

Wash your hands when finished.

Wear goggles and an apron.

Waste Disposal

Your teacher will provide a waste container for the solutions used in this experiment.

Preliminary Tests

Obtain an unknown from the instructor.

Most tests will require that you use a solution of your unknown. To prepare the unknown solution for testing, dissolve about 0.2 g of the unknown solid in 25 mL of water.

Run the preliminary tests on both the known stock solution and your unknown.

Use a fresh unknown sample for each test unless directed otherwise.

Treatment with $AgNO_3$

Place 10 drops of the solution to be tested (unknown or known which contains all ions except NO_3^-) into a small test tube and add 10 drops of 0.2 M $AgNO_3$. Stir well to mix. Centrifuge and decant. You may discard the supernate into the waste container.

NO_3^- is a strong oxidizing agent and can not be in contact with I-, S^{2-}, and SO_3^{2-} in the known solution since it would oxidize these ions.

Wash the precipitate with several drops of cold water. Centrifuge and discard the washings.

Add 10-15 drops of 3.0 M HNO_3 to the precipitate. Record the color of the precipitate. Centrifuge and save the supernatant.

To the supernatant, add concentrated NH3 in the hood until just basic. Then add dilute acetic acid till just acidic. Finally, add 8-10 drops of 0.20 M $AgNO_3$. Record your observations.

Dispose of the sample in the waste container.

Treatment with $BaCl_2$

Place 5 drops of the solution to be tested (unknown or known which contains all ions except NO_3^-) into a clean small test tube. Add 5 drops of 0.20 M $BaCl_2$. Record your observations. Centrifuge and decant. Discard the supernatant in the waste container.

Add 10 drops of 2.0 M HCl. Stir well and warm. Record your observations. A white precipitate remaining proves the presence of sulfate.

Dispose of the sample in the waste container.

Treatment with H_2SO_4

Place 5 drops of the solution to be tested (unknown or known which contains all ions except NO_3^-) into a clean small test tube. Add 5 drops of 0.20 M $BaCl_2$. Record your observations. Centrifuge and decant. Discard the supernatant in the waste container.

Add 10 drops of 2.0 M HCl. Stir well and warm. Record your observations. A white precipitate remaining proves the presence of sulfate.

Dispose of the sample in the waste container.

Treatment with H_2SO_4

For this test, test only the unknown.

Place approximately 0.050 g of the solid salt in a small test tube. In the hood, add 2 drops of 2.0 M H_2SO_4. Examine the mixture for evidence of the evolution of gases. If no gases are observed, carefully heat the mixture in the hot water bath. If still no gases are evolved, then CO_3^{-2}, S^{-2}, and SO_3^{-2} ions are absent. If a gas is evolved, note its odor and color. Below are the properties of some of the gases that possibly formed.

- CO_2 (carbon dioxide) is colorless and odorless.
- H_2S (hydrogen sulfide) is colorless and smells like rotten eggs.
- SO_2 (sulfur dioxide) is colorless and smells like burning sulfur.

Confirmatory Tests

Before, you begin the confirmatory tests, you will need to analyze the results of the preliminary tests in order to eliminate some of the anions as possibilities. Only run the confirmatory tests on anions that are still in question.

The confirmatory test procedures should be done using both your unknown and a sample known to contain the ion under investigation. Known solutions can be made by dissolving a small amount of the sodium salt of the anion- for example sodium nitrate ($NaNO_3$) for the nitrate test, sodium chloride (NaCl) for the chloride test etc in 1-2 ml of water. Be sure all of the solid has dissolved before starting the test.

Use a fresh unknown sample for each test unless directed otherwise. To prepare a solution of your unknown, place 0.2 grams of the unknown in 25 mL of water.

Test for Sulfide lon, S^{2-} To 1 mL of test solution (known or unknown solution), add 6-8 drops of 6.0 M HCl. Note any odor of H_2S (rotten eggs). Place a strip of moistened lead acetate paper over the mouth of the test tube and heat the test tube in a water bath for several minutes. A darkening of the lead acetate paper as it forms PbS confirms the presence of the S^{2-} ion.

Note: If S^{2-} is found to be present, SO_3^{2-} cannot be present since S^{2-} reduces SO_3^{2-} to S. Therefore the SO_3^{2-} test may be omitted. Note also the modification of the PO_4^{3-} test when S^{2-} is present.

Test for Sulfate Ion, SO_4^{2-} To 2 mL of the test solution (known or unknown solution) add 6.0 M HCI drop by drop until the solution is slightly acidic. If this is a solution containing the unknown, note any gas-formation and carefully check for a sharp odor of SO_2 gas. If an odor is noted, the unknown likely has sulfite ion, SO_3^{2-}. Continue with the sulfate test by adding 1 mL of 0.1 M $BaCl_2$ solution, or more as needed to complete any precipitation. A white precipitate of BaSO4 proves the presence of SO_4^{2-}.

Test for Sulfite Ion, SO_3^{2-} continuing with the sample used the sulfate test above. Centrifuge the tube to obtain a clear filtrate. Add a drop or two of 0.1 M $BaCl_2$ to be sure all of the SO_4^{2-}. has precipitated. Centrifuge again if more BaSO4 formed in the solution. Decant the filtrate into a new test tube. To the clear solution, add a few drops of hydrogen peroxide to oxidize any SO_3^{2-} to SO_4^{2-}. A white precipitate that forms in the solution proves the presence of SO_4^{2-} as shown in the reaction below:

$$H_2O_2 + Na_2SO_3 \rightarrow Na_2SO_4 + H_2O$$

Test for Phosphate PO_4^{3-} Obtain 10 drops of the sample to be tested (known or unknown). Add 5 drops of 0.5 M (NH_4) $3MoO_4$ (ammonium molybdate) and with 5 drops 6.0 M HNO_3. A yellow precipitate of $(NH_4)3PO_4$. $12MoO_3$, indicates the presence of phosphate. If a precipitate does not form immediately, heat in a water bath for a few minutes.

The balanced reaction for this test is:

$$3NH_4^+ + 12MoO_4^{2-} + H_3PO_4 + 21H^+ \rightarrow (NH_3)3PO_4.12MoO_{3(s)} + 12H_2O$$

If your sample contains sulfide ion (S^{2-}), you must first remove it since it interferes with the phosphate test. To remove the sulfide ion, begin the test by add HCl dropwise to your 10 drops of sample to make it distinctly acidic and heat with stirring for 3 minutes to remove sulfide as H_2S gas. Continue the test by adding the ammonium molybdate and nitric acid.

Test for Carbonate CO_3^{2-}, place a small (pea sized) amount of the solid to be tested into a small test tube. Dip a Nichrome wire loop into $Ba(OH)_2$ solution so that some of the $Ba(OH)_2$ solution adheres to the wire. Place the wire with the suspended $Ba(OH)_2$ just down into the test tube containing the sample to be tested. It should not be in contact with the solid.

Add several drops of 2.0 M HCl to the test tube. The rapid evolution of tiny bubbles suggests the presence of carbonate. A cloudy precipitate of $BaCO_3$ formed in the drop of $Ba(OH)_2$ confirms that presence of carbonate.

If SO_3^{2-} is present, is may interfere with the carbonate test as it forms SO_2 gas upon addition of acid and will precipitate with $Ba(OH)_2$ to form $BaSO_3$. To remove SO_3^{2-}, begin the carbonate test by addition an equal amount of sodium peroxide (Na_2O_2) to your solid sample. Then add 5 drops of water, stir, and continue procedure using the $Ba(OH)_2$ and HCl.

Test for Chloride ion, to a 2 mL portion of the test solution (known or unknown) add a few drops of 6 M HNO_3 as needed to make the solution slightly acidic (Test with litmus paper). If you sample contains sulfide, it must be removed by boiling the solution a moment. The free sulfur formed does not interfere.

Add 1 mL of 0.1 M $AgNO_3$. If you are testing the unknown and no precipitate forms at this point, the lack of precipitate proves the absence of CI^-, Brand I^-. If a precipitate forms, note the color of the precipitate. AgCl is white, AgBr is cream colored, and AgI is yellow. The color may be a clue to the identity of the anion, but is not always reliable since the colors formed are similar.

Centrifuge the mixture: Test the clear filtrate with 1-2 drops of 0.1 M AgNO3 for complete precipitation. If the filtrate turns cloudy, centrifuge again. Discard the filtrate and keep the precipitate. Wash the precipitate with by adding 10 drops of distilled water to remove excess acid and silver ion. Discard the washings. To the precipitate that remains add 3 mL of distilled water, 4 drops of 6 M NH_4OH, and 0.5 mL of 0.1 M $AgNO_3$. (The proportions are important because we wish to dissolve only the AgCl from any mixture of AgCl, AgBr, AgI and Ag_2S.) The soluble ions $Ag(NH_3)_2^+$ and Clwill form if AgCl is present. These ions will be present in the solution. Shake the mixture well and centrifuge. Transfer the clear solution to a clean test tube and acidify with 6 M HNO_3. A white precipitate of AgCl confirms the presence of Cl^-.

Test for Iodide and Bromide ions, I^- and Br^-. The complete separation and testing for Iand Bris based upon the selective oxidation of the I^- ion and Br^- ion and their color in cyclohexane. The color of the I_2 (purple) is so dark it will hide the color of Br_2 (brown) if it is also present. Therefore, in order to simplify this experiment, the unknown will not contain both Br and I.$^-$

To 2 mL of the test solution, add 6.0 M HCl to make the solution acidic. If you are testing the unknown solution and either S^{2-} or SO_3^{2-} has been found in the unknown, boil the solution to remove the ion. Now add 1 to 2 mL of chlorine water and 2 mL of cyclohexane, stopper the tube, and shake. A purple color indicates I^- is present; a brown (amber) color indicates Br^- is present.

Test for Acetate Ion

Place a small amount of the solid to be tested for acetate ion in a small test tube. Add 3-4 drops of concentrated (18 M) H_2SO_4 and mix thoroughly. Add 4-5 drops of ethanol (ethyl alcohol) and mix thoroughly. Heat in a boiling water bath for 1 minute. Carefully smell the order of the escaping fumes by wafting. A fruity odor of ethyl acetate (similar to juicy fruit gum) escaping proves the presence of acetate ion. Note that if nitrate is present, it forms ethyl nitrate which has a similar order but somewhat more sweet smelling.

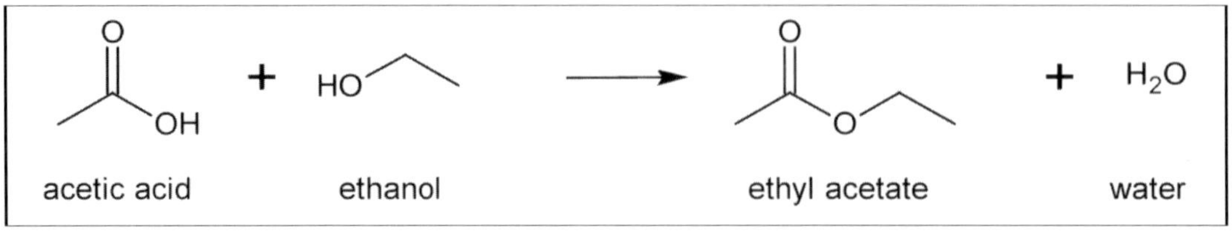

Test for nitrate ions, NO_3^- Iodide and bromide ions (I^- and Br^-) interfere with the nitrate test so they must be removed first if they are present.

If Br^- or Iion is present, treat 10 drops of the test solution with 1 ml of saturated lead acetate Ag-$C_2H_3O_2$ solution to precipitate out AgBr or AgI. Centrifuge and use the clear supernatant for the nitrate test described below.

If neither Br^- nor I^- is present, use 10 drops of the test solution. Add 5 drops freshly prepared $FeSO_4$ solution (you must make this solution yourself by dissolving a small amount solid $FeSO_4$ in 2 mL H_2O) and mix. Acidify with 3.0 M H_2SO_4, mixing thoroughly. Incline the test tube at a 450 angle and gently pour about 1 mL of concentrated H_2SO_4 (18 M) down the side of the test tube.

The concentrated H_2SO_4 will form a layer in the bottom of the tube because of its greater density. Avoid mixing. Let the test tube stand for 3-5 minutes. Look for the presence of a brown ring at the interface of the concentrated H_2SO_4 layer and the test solution layer. The brown ring is $Fe(NO)^{2+}$ and indicates the presence of NO_3^-. A faint test may be observed more easily by holding the test tube against a white piece of paper. The reaction you have observed occurs in two steps:

$$NO_3^- + 3Fe^{2+} + 4H^+ \rightarrow NO + 3Fe^{3+} + 2H_2O$$

$$Fe^{2+} + NO \rightarrow [Fe(NO)^{2+}]$$

Nitrate Test

A nitrate test is a chemical test used to determine the presence of nitrate ion in solution. Testing for the presence of nitrate via wet chemistry is generally difficult compared with testing for other anions, as almost all nitrates are soluble in water. In contrast, many common ions give insoluble salts, e.g. halides precipitate with silver, and sulfate precipitate with barium.

The nitrate anion is an oxidizer, and many tests for the nitrate anion are based on this property. Unfortunately, other oxidants present in the analyte may interfere and give erroneous results.

Ring Test

A common nitrate test, known as the brown ring test can be performed by adding iron(II) sulfate to a solution of a nitrate, then slowly adding concentrated sulfuric acid such that the acid forms a layer below the aqueous solution. A brown ring will form at the junction of the two layers, indicating the presence of the nitrate ion. Note that the presence of nitrite ions will interfere with this test.

Nitrate Test by using Brown Ring Test.

The overall reaction is the reduction of the nitrate ion by iron(II) which is oxidised to iron(III) and formation of a nitrosonium complex where nitric oxide is reduced to NO^-.

$$2HNO_3 + 3H_2SO_4 + 6FeSO_4 \rightarrow 3Fe_2(SO_4)_3 + 2NO + 4H_2O$$

$$[Fe(H_2O)_6]SO_4 + NO = [Fe(H_2O)_5(NO)]\,SO_4 + H_2O$$

This test is sensitive up to 2.5 micrograms and a concentration of 1 in 25,000 parts.

Devarda's Test

Devarda's alloy (Copper/Aluminium/Zinc) is a reducing agent. When reacted with nitrate in sodium hydroxide solution, ammonia is liberated. The ammonia formed may be detected by its characteristic odor, and by damp red litmus paper's turning blue, signalling that it is an alkali — very few gases other than ammonia evolved from wet chemistry are alkaline.

$$3NO_3^- + 8Al + 5OH^- + 18H_2O \rightarrow 3NH_3 + 8[Al(OH)_4]^-$$

Aluminium is the reducing agent in this reaction that will occur.

Diphenylamine Test

Diphenylamine may be used as a wet chemical test for the presence of the nitrate ion. In this test, a solution of diphenylamine and ammonium chloride in sulfuric acid is used. In the presence of nitrates, diphenylamine is oxidized, giving a blue coloration. This reaction has been used to test for organic nitrates as well, and has found use in gunshot residue kits detecting nitroglycerine and nitrocellulose.

Copper Turnings Test

The nitrate ion can easily be identified by heating copper turnings along with concentrated sulphuric acid. Brown, pungent effervescence is observed which turns blue litmus red.

Here sulphuric acid reacts with the nitrate ion to form nitric acid. Nitric acid then reacts with the copper turnings to form nitric oxide. Nitric oxide is thus oxidised to nitrogen dioxide.

$$Cu + 4HNO_3 \rightarrow Cu(NO_3)_2 + 2NO_2 + 2H_2O$$

Nitrite Test

A nitrite test is a chemical test used to determine the presence of nitrite ion in solution.

Chemical Method

A common nitrite test can be performed by adding 4 M sulfuric acid to the sample until acidic, and then adding 0.1 M iron(II) sulfate to the solution. A positive test for nitrite is indicated by a dark brown solution, arising from the iron-nitric oxide complex ion.

Urinary Nitrite Test

A nitrite test is a standard component of a urinary test strip. A positive test for nitrites in the urine is called nitrituria.

This test is commonly used in diagnosing urinary tract infections (UTIs). A positive nitrite test indicates that the cause of the UTI is a gram negative organism, most commonly *Escherichia coli*. The reason for nitrites' existence in the presence of a UTI is due to a bacterial conversion of endogenous nitrates to nitrites. This may be a sign of infection. However, other parameters, such as leukocyte esterase, urine white blood cell count, and symptoms such as dysuria, urinary urgency, fevers, and chills must be correlated to diagnose an infection

False negative nitrite tests in urinary tract infections occur in cases with a low colony forming unit count, or in recently voided or dilute urine. In addition, a nitrite test does not detect organisms unable to reduce nitrate to nitrite, such as enterococci, staphylococci (*Staphylococcus saprophyticus*), Acinetobacter, or adenovirus.

Qualitative Analysis of Cations

Cations are usually classified into six groups. Each group has a common reagent that can be used to separate them from the solution. Because cationic analysis is based on the solubility products of the ions, meaningful results can be obtained only if separation is performed in a specified sequence. This is due to the fact that some ions of one group may also react with the reagent of another group. For example, both Ba^{2+} and Sr^{2+} will react with the SO_4^{2-} ion to form a solid. Therefore, mathematical calculations should be done before choosing the SO_4^{2-} ion for selective precipitation in a solution that may contain both Ba^{2+} and Sr^{2+}.

Preliminary Tests

Some preliminary tests needs to be done before doing the analysis of cations.

(a) Physical Appearance: The physical examination of the unknown salt involves the study of color, smell and density. The test is not much reliable, but certainly helpful in identifying some colored cations. Characteristic smell helps to identify some ions like ammonium ion.

(b) Charcoal Cavity Test: This test is based on the fact that metallic carbonates when heated in a charcoal cavity decomposes to give corresponding oxides. The oxides appear as colored incrustation or residue in the cavity. In certain cases, the oxides formed partially undergo reduction to the metallic state producing metallic beads or scales.

Examples:

$$a)\ ZnSo_4 + Na_2Co \rightarrow ZnCO_3 + Na_2SO_4$$

$$ZnCO_3 \rightarrow \underset{\substack{\text{yellow when hot} \\ \text{white when cold}}}{ZnO} + CO_2 \uparrow$$

$$b)\ Pb(NO_3)_2 + Na_2\,CO_3 \rightarrow Pb\,CO_3 + 2\,NaNO_3$$

$$Pb\,CO_3 \rightarrow PbO + CO_2 \uparrow$$

$$PbO + C \rightarrow \underset{\text{Bead}}{Pb} + CO$$

$$c)\ CuSO_4 + Na_2CO_3 \rightarrow CuCO_3 + Na_2SO_4$$

$$CuCO_3 \rightarrow CuO + CO_2 \uparrow$$

$$CuO + C \rightarrow \underset{(\text{Reddish scales})}{Cu} + CO$$

(c) Cobalt Nitrate Test (Ash test): This test is applied to those salts that leave white residue in the charcoal cavity test. This test is based on the fact that metallic carbonates when heated in a charcoal cavity decomposes to give corresponding oxides.

The test is based on the fact that cobalt nitrate decomposes on heating to give cobalt oxide, CoO. This combines with the metallic-oxides present as white residue in the charcoal cavity forming colored compounds.

$$2\,Co(NO_3)_2 \xrightarrow{\Delta} 2\,CoO + 4\,NO_2 + O_2$$

For example, when a magnesium salt undergoes charcoal cavity test, a while residue of MgO is left behind. This on treatment with cobalt nitrate and on subsequent heating forms a double salt of the formula MgO.CoO, which is pink in color.

$$MgO + CoO \rightarrow MgO.CoO(\text{Pink})$$

Other examples are:

$$ZnO + CoO \rightarrow ZnO.CoO(\text{Green})$$

$$AI_2O_3 + CoO \rightarrow AI_2O_3.CoO(\text{Blue})$$

In addition to metallic oxides, phosphates and borates also react with cobalt oxide to form $Co_3(PO_4)_2$ and $Co_3(BO_3)_2$ that are blue in color.

(d) Flame Test: Certain salts on reacting with conc. HCl forms their chloride, that are volatile in non-luminous flame. Their vapours impart characteristic color to the flame. This color can give reliable information of the presence of certain cations. For proceeding to this test, the paste of the mixture with conc. HCl is introduced into the flame using a platinum wire.

(e) Borax Bead Test: Borax, $Na_2B_4O_7.10H_2O$, on heating gets fused and loses water of crystallisation. It swells up into a fluffy white porous mass which then melts into a colorless liquid that later forms a clear transparent glassy bead consisting of boric anhydride and sodium metaborate.

$$Na_2B_4O_7.10H_2O \rightarrow Na_2B_4O_7 + 10H_2O\uparrow$$

$$Na_2B_4O_7 \rightarrow \underset{\text{Boric anhydride}}{B_2O_3} + \underset{\text{Sodium meta borate}}{2NaBO_2}$$

Boric anhydride is non-volatile. When it is made to react with colored metallic salt, a characteristic colored bead of metal metaborate is formed. In those cases where different colored beads are obtained in oxidising and reducing flames, metaborates in various oxidation states of metals are formed. For example, in oxidising flame, copper forms blue copper metaborate.

$$Na_2B_4O_7 + CuSO_4 \rightarrow 2NaBO_2 + \underset{\text{(Blue)}}{Cu(BO_2)_2} + SO_3$$

In reducing flame, cupric metaborate is reduced to metallic copper, which is red and opaque.

$$2Cu(BO_2)_2 + 4NaBO_2 + 2C \rightarrow \underset{\text{(Red opaque)}}{2Cu} + 2Na_2B_4O_7 + 2CO$$

Identification of Cations: Basic Radicals

Analysis of Group Zero

This group includes ammonium ion (NH_4^+).

Chemical reactions involved in Group Zero analysis:

(a) Sodium hydroxide test: When ammonium salt is heated with conc. NaOH, ammonia gas is evolved which gives white fumes with dil.HCl due to the formation of NH_4Cl.

$$NH_4CI + NaOH \xrightarrow{\Delta} NaCI + H_2O + \underset{\text{Ammonia}}{NH_3}\uparrow$$

$$NH_3 + HCI \rightarrow \underset{\substack{\text{Ammonium chloride} \\ \text{(White fumer)}}}{NH_4CI}$$

(b) Nessler's reagent test: Tha ammonia gas formed by the reaction of ammonium ions with NaOH reacts with Nessler's reagent to form a brown precipitate of $H_2N.HgO.HgI$.

$$\underset{\text{Nessler's reagent}}{2K_2[HgI_4]} + NH_3 + 3KOH \rightarrow \underset{\text{Brown precipitate}}{H_2N.HgO.HgI}\downarrow + 7KI + 2H_2O$$

Analysis of Group I

This group includes Pb^{2+}, Ag^{+}, Hg_2^{2+}. Here we shall study only Pb^{2+}. Group reagent for this group is dil. HCl.

Chemical reactions involved in Group I analysis:

The addition of HCl to the solution will precipitate Pb^{2+} as lead chloride which is soluble in hot water. On cooling, the precipitate settle down as $PbCl_2$ which is less soluble in cold water.

$$Pb(NO_3) + 2HCI \rightarrow \underset{\substack{\text{Lead chloride} \\ \text{(white precipitate)}}}{PbCI_2 \downarrow} + 2HNO_3$$

Confirmatory tests of Lead (II) ion (Pb^{2+}):

(a) Potassium iodide test: Lead chloride (formed by the reaction of lead salts with dil. HCl) solution in hot water reacts with potassium iodide solution to form yellow precipitate of lead iodide.

$$\underset{\text{(Hot solution)}}{PbCI_2} + 2KI \rightarrow \underset{\text{yellow precipitate}}{PbCI_2 \downarrow} + 2KCI$$

(b) Potassium chromate test: Lead chloride (formed by the reaction of lead salts with dil. HCl) solution in hot water reacts with solution of potassium chromate to form yellow precipitate of lead chromate.

$$\underset{\text{(Hot solution)}}{PbCI_2} + K_2CrO_4 \rightarrow \underset{\text{yellow precipitate}}{PbCrO_4 \downarrow} + 2KCI$$

Analysis of Group II

This group includes Pb^{2+} and Cu^{2+} in IIA Group and As^{3+} in IIB Group.

Chemical reactions involved in Group II Analysis:

Passing of H_2S gas through the acidified original solution will precipitate the radicals Pb^{2+},Cu^{2+},As^{3+} as their sulphides.

$$PbCI_2 + H_2S \rightarrow 2HCI + PbS \downarrow \text{ (Black)}$$

$$CuCl_2 + H_2S \rightarrow 2HCI + CuS \downarrow \text{ (Black)}$$

$$2AsCl_3 + 3H_2S \rightarrow 6HCI + As_2S_3 \downarrow \text{(yellow)}$$

Confirmatory tests for Lead (II) ion (Pb^{2+})

Black precipitate of PbS formed in the group analysis dissolves in 50% nitric acid due to the formation of soluble lead nitrate. On adding sulphuric acid to the soluble lead nitrate, lead sulphate precipitates.

$$3PbS + 8HNO_3 \rightarrow \underset{\substack{\text{Lead nitrate} \\ \text{(Soluble)}}}{3Pb(NO_3)_2} + 4H_2O + 2NO + 3S$$

$$Pb(NO_3)_2 + H_2SO_4 \rightarrow 2HNO_3 + \underset{\substack{\text{Lead sulphate} \\ \text{(white)}}}{PbSO_4 \downarrow}$$

(a) Potassium iodide test: The soluble lead nitrate (formed by the reaction between PbS and 50% conc. HNO_3) reacts with potassium iodide to form yellow precipitate of lead iodide.

$$Pb(NO_3)_2 + 2KI \rightarrow 2KNO_3 + \underset{\substack{\text{Lead iodide} \\ \text{(yellow)}}}{PbI_2} \downarrow$$

(b) Potassium chromate test: The soluble lead nitrate (formed by the reaction between PbS and 50% conc. HNO_3) reacts with potassium chromate to form yellow precipitate of lead chromate.

$$Pb(NO_3)_2 + K_2CrO_4 \rightarrow 2KNO_3 + \underset{\substack{\text{Lead chromate} \\ \text{(yellow)}}}{PbCrO_4} \downarrow$$

Confirmatory tests for Copper (II) ion (Cu^{2+}):

Black precipitate of CuS formed in the group analysis dissolves in 50% nitric acid and a blue solution is obtained on addition excess of NH_4OH.

$$3CuS + 8HNO_3 \rightarrow 3Cu(NO_3)_2 + 4H_2O + 2NO + 3S$$

$$Cu(NO_3)_2 + 4NH_4OH \rightarrow \underset{\text{(Blue solution)}}{[Cu(NO_3)_4][NO_3]_2} + 4H_2O$$

(a) Potassium ferrocyanide test: The above blue solution gives a chocolate brown precipitate with potassium ferrocyanide solurion.

[üüü(üüüüü)üüüü]üüü 4+ 3 → 4+ 3 4
(Blue solution)

$$2CuSO_4 + \underset{\text{Potassium ferrocyanide}}{K_4[Fe(CN)_6]} \rightarrow \underset{\text{(Chocolate brown precipitate)}}{Cu_2[Fe(CN)_6]} \downarrow + 2K_2SO_4$$

(b) Potassium iodide test: In this test, the white precipitate is due to the formation of cuprous iodide and the brown color of the solution is due to liberation of iodine.

$$2CuSO_4 + 4KI \rightarrow \underset{\text{(White precipitate)}}{Cu_2I_2} + \underset{\text{(Brown colouration)}}{I_2} + 2K_2SO_4$$

Confirmatory tests of Arsenic (III) ion (As^{3+}): The yelllow residue of As_2S_3 formed in the group analysis is dissolved in conc. HNO_3 forming arsenic acid.

$$As_2S_3 + 10HNO_3 \rightarrow 2\underset{\substack{\text{Arsenic acid} \\ \text{(soluble)}}}{H_3AsO_4} + 10NO_2 + 3S + 2H_2O$$

(a) Ammonium molybdate test: Arsenic acid (formed by dissolving As_2S_3 in conc. HNO_3) reacts with ammonium molybdate to form yellow precipitate of ammonium arseno molybdate.

$$\underset{\text{Arsenic acid}}{H_3AsO_4} + 12(NH_4)_2MoO_4 + 21HNO_3 \rightarrow \underset{\substack{\text{Ammonium arseno molybdate} \\ \text{(yellow precipitate)}}}{(NH_4)_3AsO_4.12MoO_3} \downarrow + 21NH_4NO_3 + 12H_2O$$

(b) Magnesia mixture test: Arsenic acid (formed by dissolving As_2S_3 in conc. HNO_3) reacts with magnesia mixture to form a white precipitate of $Mg(NH_4)_2AsO_4$.

$$\underset{\text{Arsenic acid}}{H_3AsO_4} + MgSO_4 + 3\,NH_4OH \rightarrow \underset{\text{(White precipitate)}}{Mg(NH_4)_2\,AsO_4} + (NH_4)_2\,SO_4 + 3\,H_2O$$

Analysis of Group III

The cations present in this group are Fe^{2+}, Fe^{3+}, Cr^{3+} and Al^{3+}. We will look at only Fe^{2+}/Fe^{3+} and Al^{3+}.

Chemical Reactions Involved Group III Analysis

The cations in this group are precipitated as hydroxides by adding ammonium hydroxide in presence of ammonium chloride. Thus, group reagent for this group is NH_4OH in the presence of NH_4Cl.

$$FeCl_3 + 3\,NH_4OH \rightarrow 3\,NH_4Cl + \underset{\substack{\text{Ferric hydroxide}\\ \text{(Reddish brown)}}}{Fe(OH)_3} \downarrow$$

$$AlCl_3 + 3\,NH_4OH \rightarrow 3\,NH_4Cl + \underset{\substack{\text{Aluminium hydroxide}\\ \text{(white)}}}{Al(OH)_3} \downarrow$$

Confirmatory tests of Ferric ion (Fe^{3+}): The reddish brown precipitate of ferric hydroxide is dissolved in HCl due to the formation of soluble ferric chloride.

$$Fe(OH)_3 + 3\,HCI \rightarrow \underset{\text{Ferric chloride}}{FeCl_3} + 3\,H_2O$$

(a) Potassium ferrocyanide test: Ferric chloride (formed by dissolving ferric hydroxide in HCl) reacts with potassium ferrocyanide to form prussian blue colored Ferric ferrocyanide.

$$4\,FeCl_3 + 3\,K_4\left[Fe(CN)_6\right] \rightarrow 12\,KCI + \underset{\substack{\text{Ferric ferrocy anide}\\ \text{(Prussian blue)}}}{Fe_4\left[Fe(CN)_6\right]_3}$$

(b) Potassium sulphocyanide test: Ferric chloride (formed by dissolving ferric hydroxide in HCl) reacts with potassium sulphocyanide to form blood red colored Ferric sulphocyanide.

$$FeCl_3 + KCNS \rightarrow 3\,KCI + \underset{\substack{\text{Ferric sulphocy anide}\\ \text{(Blood red colour)}}}{Fe(CNS)_3}$$

Confirmatory tests of Aluminium (III) ion (Al^{3+}):

(a) Lake test: Aluminium hydroxide formed in the group analysis dissolves in dil. HCl to form soluble aluminium chloride. The aluminium chloride thus formed reacts with ammonium hydroxide to reform aluminium hydroxide. Blue color of litmus solution is adosrbed on this precipitate.

$$Al(OH_3 + 3\,HCI \rightarrow AlCl_3 + 3\,H_2O \text{ (dissolution)}$$

$$AlCl_3 + 3\,NH_4OH \rightarrow 3\,NH_4Cl + \underset{\substack{\text{Aluminium hydroxide}\\ \text{(Blue colour ad sorbed on this preciptate)}}}{Al(OH)_3} \downarrow$$

(b) Charcoal cavity/Cobalt nitrate test: In this test, aluminium oxide produced in the charcoal cavity test reacts with CoO in cobalt nitrate test to produces a blue mass due to the formation of $Al_2O_3.CoO$.

$$Al_2O_3 + CoO \rightarrow Al_2O_3.CoO(\text{Blue mass})$$

Analysis of Group IV

The radicals present in this group are CO^{2+}, Ni^{2+}, Mn^{2+} and Zn^{2+}. These are precipitated as sulphide by passing H_2S gas through the ammonical solution of the salt. The group reagent for this group is H_2S gas in the presence of NH_4Cl and NH_4OH.

Chemical reaction involved in Group IV analysis:

Passing of H_2S gas through the group III solution will precipitate the radicals Co^{2+}, Ni^{2+}, Mn^{2+} and Zn^{2+} as their sulphides. Formation of black ppt. (CoS or NiS) indicates cobalt or nickel. Formation of buff-colored ppt. (MnS) indicates manganese and dirty white ppt. (ZnS) indicates zinc.

$$Co(OH)_2 + H_2S \rightarrow 2H_2O + CoS\downarrow(\text{Black})$$
$$Ni(OH)_2 + H_2S \rightarrow 2H_2O + NiS\downarrow(\text{Black})$$
$$Zn(OH)_2 + H_2S \rightarrow 2H_2O + ZnS\downarrow(\text{Dull white})$$
$$Mn(OH)_2 + H_2S \rightarrow 2H_2O + MnS\downarrow(\text{Buff or flesh colour})$$

Confirmatory tests of Cobalt (II) ion (Co^{2+}):

(a) Potassium nitrite test: Cobalt ion reacts with potassium nitrite in presence of acetic acid to form yellow precipitate of Potassium cobalti nitrite.

$$CoCl_2 + 2KNO_2 \rightarrow 2KCl + \underset{\text{Cobaltous nitrite}}{Co(NO_2)_2}$$
$$KNO_2 + CH_3COOH \rightarrow CH_3COOK + HNO_2$$
$$Co(NO_2)_2 + 2HNO_2 \rightarrow \underset{\text{Cobalti nitrite}}{CO(NO_2)_3} + H_2O + NO$$
$$Co(NO_2)_2 + 3KNO_2 \rightarrow \underset{\substack{\text{Potassium cobalti nitrite}\\ \text{(yellow precipitate)}}}{K_3[Co(NO_2)_6]}$$

(b) Ammonium thiocyanate ether test: On addition of ether and a crystal of ammonium thiocyanate, a blue color is obtained in the ethereal layer due to the formation of ammonium cobalti thiocyanate.

$$CoCl_2 + 4NH_4CNS \rightarrow \underset{\text{Ammonium cobalti thiocyanate}}{(NH_4)_2[Co(CNS)_4]} + 2NH_4CI$$

(c) Borax bead test: In this test Co^{2+} ion produces blue beads in both oxidising and reducing flames.

Confirmatory tests of Nickel (II) ion (Ni^{2+}):

(a) Dimethyl glyoxime test: Ni^{2+} ions react with dimethyl glyoxime to form bright rose-red colored Nickel-Dimethyl glyoxime complex, $Ni(dmgH)_2$.

(b) Sodium hydroxide-bromine water test: Ni^{2+} ions react with excess of NaOH and bromine water to form a black precipitate of Nickelic hydroxide.

$$NiCl_2 + 2\,NaOH \rightarrow 2\,NaCl + \underset{\text{(Green precipitate)}}{Ni(OH)_2 \downarrow}$$

$$Br_2 + H_2O \rightarrow 2\,HBr + O$$

$$2\,Ni(OH)_2 + H_2O + O \rightarrow \underset{\substack{\text{Nick elic hydroxide} \\ \text{(Black precipitate)}}}{2\,Ni(OH)_3 \downarrow}$$

(c) Borax bead test: Ni^{2+} ions produce brown bead in oxidising and grey bead in reducing flames.

Confirmatory tests of Manganese (II) ion (Mn^{2+}):

Manganese sulphide formed in the group analysis dissolves in dil.HCl forming manganese chloride, and H_2S is boiled off.

$$MnS + 2\,HCl \rightarrow MnCl_2 + H_2S \uparrow$$

(a) Sodium hydroxide-bromine water test: The manganese chloride (formed by dissolving MnS in HCl) reacts with excess of NaOH to form white precipitate of manganese hydroxide.

$$MnCl_2 + 2\,NaOH \rightarrow \underset{\substack{\text{Manganese hydroxide} \\ \text{(White precipitate)}}}{Mn(OH)_2 \downarrow} + 2\,NaCl$$

The white precipitate of manganese hydroxide turns brown on addition of bromine water due to its oxidation to brown manganic hydroxide $MnO(OH)_2$.

$$Br_2 + H_2O \rightarrow 2\,HBr + [O]$$

$$Mn(OH)_2 + [O] \rightarrow \underset{\substack{\text{Manganic hydroxide} \\ \text{(Brown precipitate)}}}{MnO(OH)_2 \downarrow}$$

(b) Lead peroxide test: Manganese sulphide formed in the group analysis reacts with conc. HNO_3 and lead peroxide to form a pink colored solution, due to the formation of $HMnO_4$.

$$MnS + 2\,HNO_3 \rightarrow Mn(NO_3)_2 + H_2S$$

$$2\,Mn(NO_3)_2 + 5\,PbO_2 + 6\,HNO_3 \rightarrow \underset{\text{Pink solution}}{2\,HMnO_4} + 5\,Pb(NO_3)_2 + 2\,H_2O$$

(c) Borax bead test: Mn^{2+} ions produce pinkish beads in oxidising and colorless beads in reducing flames.

Confirmatory tests of Zinc (II) ion (Zn^{2+}):

The white precipitate of ZnS formed in the group analysis dissolves in dil. HCl forming zinc chloride, and H_2S is boiled off.

$$ZnS + 2\,HCl \rightarrow ZnCl_2 + H_2S \uparrow$$

(a) Sodium hydroxide test: Zinc chloride (formed by dissolving ZnS in dil. HCl) reacts with sodium hydroxide to form white precipitate of zinc hydroxide.

$$ZnCl_2 + 2\,NaOH \rightarrow \underset{\substack{\text{Zinc hydroxide} \\ \text{(White precipitate)}}}{Zn(OH)_2} \downarrow + 2\,NaCl$$

(b) Potassium ferrocyanide test: Zinc chloride (formed by dissolving ZnS in dil. HCl) reacts with potassium ferrocyanide to form white or bluish white precipitate of zinc ferrocyanide.

$$2\,ZnCl_2 + K_4\,[Fe\,(CN)_6] \rightarrow \underset{\substack{\text{Zins ferrocyanide} \\ \text{(White or Bluish white)}}}{Zn_2[Fe\,(CN)_6]} \downarrow + 4\,KCl$$

(c) Charcoal cavity/Cobalt nitrate test: In this test, zinc oxide produced in the charcoal cavity test reacts with CoO in cobalt nitrate test to produces a greenish residue due to the formation of ZnO.CoO.

$$ZnO + CoO \rightarrow ZnO.CoO \text{ (Greenish residue)}$$

Analysis of Group V

Group V consist of three radicals: Ba^{2+}, Sr^{2+} and Ca^{2+}. These cations are precipitated as their carbonates. Group reagent for this group is $(NH_4)_2CO_3$ in the presence of NH_4Cl and NH_4OH.

Chemical reaction involved in Group V analysis:

When $(NH_4)_2CO_3$ is added to salt solution containing NH_4Cl and NH_4OH, the carbonates of Ba^{2+}, Sr^{2+} and Ca^{2+} are precipitated.

$$BaCl_2 + (NH_4)_2\,CO_3 \rightarrow BaCO_3 + 2\,NH_4Cl$$
$$SrCl_2 + (NH_4)_2\,CO_3 \rightarrow SrCO_3 + 2\,NH_4Cl$$
$$CaCl_2 + (NH_4)_2\,CO_3 \rightarrow CaCO_3 + 2\,NH_4Cl$$

Confirmation of Barium (II) ion (Ba^{2+}):

The white precipitate of barium carbonate formed in the group analysis dissolves in hot dil. acetic acid due to the formation of soluble barium acetate.

$$BaCO_3 + 2\,CH_3COOH \rightarrow \underset{\text{Barium acetate}}{(CH_3COO)_2Ba} + CO_2 \uparrow H_2O$$

(a) Potassium chromate test: Barium acetate (formed by dissolving barium carbonate in dil. acetic acid) reacts with potassium chromate to form yellow precipitate of barium chromate.

$$(CH_3COO)_2\,Ba + K_2CrO_4 \rightarrow 2\,CH_3COOK + \underset{\substack{\text{Barium chromate} \\ \text{(Yellow precipitate)}}}{BaCrO_4} \downarrow$$

(b) Flame test: Barium imparts a grassy green color to the flame.

Confirmation of Strontium (II) ion (Sr^{2+}):

The white precipitate of strontium carbonate formed in the group analysis dissolves in hot dil. acetic acid due to the formation of soluble strontium acetate.

$$SrCO_3 + 2CH_3COOH \rightarrow \underset{\text{Strontium acetate}}{(CH_3COO)_2Sr} + CO_2 \uparrow H_2O$$

(a) Ammonium sulphate test: Strontium acetate (formed by dissolving Strontium carbonate in dil. acetic acid) reacts with ammonium sulphate to form a white precipitate o strontium sulphate.

$$(CH_3COO)_2\,Sr + (NH_4)_2\,SO_4 \rightarrow 2\,CH_3COONH_4 + \underset{\substack{\text{Strontium sulphate} \\ \text{(White precipitate)}}}{SrSO_4 \downarrow}$$

(b) Flame test: Strontium imparts a crimson red color to the flame.

Confirmation of Calcium (II) ion (Ca^{2+}):

The white precipitate of calcium carbonate formed in the group analysis dissolves in hot dil. acetic acid due to the formation of soluble calcium acetate.

$$CaCO_3 + 2CH_3COOH \rightarrow \underset{\text{Calcium acetate}}{(CH_3COO)_2Ca} + CO_2 \downarrow H_2O$$

(a) Ammonium oxalate test: Calcium acetate (formed by dissolving calcium carbonate in dil. acetic acid) reacts with ammonium oxalate to form a white precipitate o calcium oxalate.

$$(CH_3COO)_2\,Ca + (NH_4)_2\,C_2O_4 \rightarrow 2\,CH_3COONH_4 + \underset{\substack{\text{Calcium oxalate} \\ \text{(White precipitate)}}}{CaC_2O_4 \downarrow}$$

(b) Flame test: Calcium imparts brick red color to the flame.

Analysis of Group VI

Chemical reaction involved in confirmation of Mg^{2+}:

(a) Ammonium Phosphate test: Mg^{2+} ions react with ammonium phosphate in presence of NH_4Cl and NH_4OH to form white precipitate of magnesium ammonium phosphate.

$$MgCl_2 + NH_4OH + (NH_4)_2\,HPO_4 \rightarrow \underset{\substack{\text{Magnesium ammonium phosphate} \\ \text{(White precipitate)}}}{Mg(NH_4)\,PO_4 \downarrow} + 2\,NH_4\,Cl + H_2O$$

(b) Charcoal cavity/Cobalt nitrate test: In this test, magnesium oxide produced in the charcoal cavity test reacts with CoO in cobalt nitrate test to produces a pink mass due to the formation of MgO.CoO.

$$MgO + CoO \rightarrow MgO.CoO \text{ (Pink mass)}$$

References

- Qualitative-chemical-analysis, science: britannica.com, Retrieved 30 April, 2019
- Qualitative-analysis-in-chemistry-608171:thoughtco.com, Retrieved 29 June, 2019
- Sanger, michael j.; phelps, amy j.; catherine banks (2004). "simple flame test techniques using cotton swabs". Journal of chemical education. 81 (7): 969. Bibcode:2004jched..81..969s. Doi:10.1021/ed081p969
- Qualitative-analysis-of-anions: chemistrynotmystery.blogspot.com, Retrieved 14 July, 2019
- Qualitative-chemical-analysis, trident-boundless-chemistry: lumenlearning.com, Retrieved 11 January, 2019

Equilibrium Chemistry 5

- Chemical Equilibrium
- Equilibrium Chemistry
- Law of Chemical Equilibrium
- Equilibrium Constant
- Van 't Hoff Equation
- Le Chatelier's Principle

The condition in a reversible chemical reaction in which there is no net change in the amount of products and reactants is known as chemical equilibrium. Different aspects of equilibrium chemistry include law of chemical equilibrium, equilibrium constant, etc. This chapter has been carefully written to provide an easy understanding of these aspects of equilibrium chemistry.

Chemical Equilibrium

Chemical equilibrium is a condition in the course of a reversible chemical reaction in which no net change in the amounts of reactants and products occurs. A reversible chemical reaction is one in which the products, as soon as they are formed, react to produce the original reactants. At equilibrium, the two opposing reactions go on at equal rates, or velocities, hence there is no net change in the amounts of substances involved. At this point the reaction may be considered to be completed; *i.e.*, for some specified reaction condition, the maximum conversion of reactants to products has been attained.

The conditions that pertain to equilibrium may be given quantitative formulation. For example, for the reversible reaction $A \leftrightharpoons B + C$, the velocity of the reaction to the right, r_1, is given by the mathematical expression (based on the law of mass action) $r_1 = k_1(A)$, where k_1 is the reaction-rate constant and the symbol in parentheses represents the concentration of A. The velocity of the reaction to the left, r_2, is $r_2 = k_2(B)(C)$. At equilibrium, $r_1 = r_2$, therefore:

$$k_1(A)_e = k_2(B)_e(C)_e,$$

$$k_1/k_2 = (B)_e(C)_e/(A)_e = K.$$

The subscript *e* represents conditions at equilibrium. For a given reaction, at some specified condition of temperature and pressure, the ratio of the amounts of products and reactants present at equilibrium, each raised to their respective powers, is a constant, designated the equilibrium constant of the reaction and represented by the symbol *K*. The value of the equilibrium constant varies with the temperature and pressure according to the principle of Le Chatelier.

By methods of statistical mechanics and chemical thermodynamics, it can be shown that the equilibrium constant is related to the change in the thermodynamic quantity called the standard Gibbs free energy accompanying the reaction. The standard Gibbs free energy of the reaction, ΔG°, which is the difference between the sum of the standard free energies of the products and that of the reactants, is equal to the negative natural logarithm of the equilibrium constant multiplied by the so-called gas constant *R* and the absolute temperature *T*:

$$\Delta G^{\circ} = -RT \ln K.$$

The equation allows the calculation of the equilibrium constant, or the relative amounts of products and reactants present at equilibrium, from measured or derived values of standard free energies of substances.

Equilibrium Chemistry

Equilibrium chemistry is concerned with systems in chemical equilibrium. The unifying principle is that the free energy of a system at equilibrium is the minimum possible, so that the slope of the free energy with respect to the reaction coordinate is zero. This principle, applied to mixtures at equilibrium provides a definition of an equilibrium constant. Applications include acid–base, host–guest, metal–complex, solubility, partition, chromatography and redox equilibria.

Thermodynamic Equilibrium

A chemical system is said to be in equilibrium when the quantities of the chemical entities involved do not and *cannot* change in time without the application of an external influence. In this sense a system in chemical equilibrium is in a stable state. The system at chemical equilibrium will be at a constant temperature, pressure (or volume) and composition. It will be insulated from exchange of heat with the surroundings, that is, it is a closed system. A change of temperature, pressure (or volume) constitutes an external influence and the equilibrium quantities will change as a result of such a change. If there is a possibility that the composition might change, but the rate of change is negligibly slow, the system is said to be in a metastable state. The equation of chemical equilibrium can be expressed symbolically as,

$$\text{reactant(s)} \rightleftharpoons \text{product(s)}$$

The sign $\rightleftharpoons$ means "are in equilibrium with". This definition refers to macroscopic properties. Changes do occur at the microscopic level of atoms and molecules, but to such a minute extent that they are not measurable and in a balanced way so that the macroscopic quantities do not change. Chemical equilibrium is a dynamic state in which forward and backward reactions proceed at such rates that the macroscopic composition of the mixture is constant. Thus, equilibrium sign $\rightleftharpoons$ symbolizes the fact that reactions occur in both forward $\rightharpoonup$ and backward $\leftharpoondown$ directions.

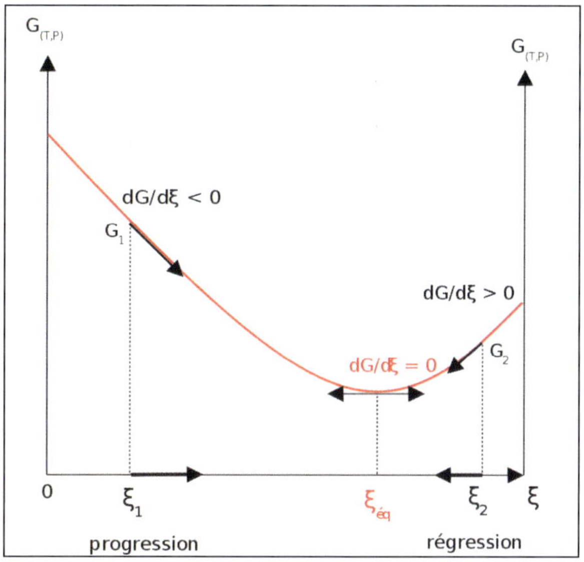

Diagram G = f (ξ)(T, p).

A steady state, on the other hand, is not necessarily an equilibrium state in the chemical sense. For example, in a radioactive decay chain the concentrations of intermediate isotopes are constant because the rate of production is equal to the rate of decay. It is not a chemical equilibrium because the decay process occurs in one direction only.

Thermodynamic equilibrium is characterized by the free energy for the whole (closed) system being a minimum. For systems at constant volume the Helmholtz free energy is minimum and for systems at constant pressure the Gibbs free energy is minimum. Thus a metastable state is one for which the free energy change between reactants and products is not minimal even though the composition does not change in time.

The existence of this minimum is due to the free energy of mixing of reactants and products being always negative. For ideal solutions the enthalpy of mixing is zero, so the minimum exists because the entropy of mixing is always positive. The slope of the reaction free energy, δG_r with respect to the reaction coordinate, ξ, is zero when the free energy is at its minimum value.

$$\delta G_r = \left(\frac{\partial G}{\partial \xi}\right)_{T,P} \quad ; \quad \delta G_r(\mathrm{Eq}) = 0$$

Acid–base Equilibria

Bronsted and Lowry characterized an acid–base equilibrium as involving a proton exchange reaction:

acid + base $\rightleftharpoons$ conjugate base + conjugate acid.

An acid is a proton donor; the proton is transferred to the base, a proton acceptor, creating a conjugate acid. For aqueous solutions of an acid HA, the base is water; the conjugate base is A^- and the conjugate acid is the solvated hydrogen ion. In solution chemistry, it is usual to use H^+ as an abbreviation for the solvated hydrogen ion, regardless of the solvent. In aqueous solution H^+ denotes a solvated hydronium ion.

The Bronsted–Lowry definition applies to other solvents, such as dimethyl sulfoxide: the solvent S acts as a base, accepting a proton and forming the conjugate acid SH^+. A broader definition of acid dissociation includes hydrolysis, in which protons are produced by the splitting of water molecules. For example, boric acid, $B(OH)_3$, acts as a weak acid, even though it is not a proton donor, because of the hydrolysis equilibrium:

$$B(OH)_3 + H_2O \rightleftharpoons B(OH)^-_4 + H^+.$$

Similarly, metal ion hydrolysis causes ions such as $[Al(H_2O)_6]^{3+}$ to behave as weak acids:

$$[Al(H_2O)_6]^{3+} \rightleftharpoons [Al(H_2O)_5(OH)]^{2+} + H^+.$$

Acid–base equilibria are important in a very wide range of applications, such as acid–base homeostasis, ocean acidification, pharmacology and analytical chemistry.

Host–guest Equilibria

A host–guest complex, also known as a donor–acceptor complex, may be formed from a Lewis base, B, and a Lewis acid, A. The host may be either a donor or an acceptor. In biochemistry host–guest complexes are known as receptor-ligand complexes; they are formed primarily by non-covalent bonding. Many host–guest complexes has 1:1 stoichiometry, but many others have more complex structures. The general equilibrium can be written as:

$$p\,A + q\,B \rightleftharpoons A_pB_q$$

The study of these complexes is important for supramolecular chemistry and molecular recognition. The objective of these studies is often to find systems with a high binding selectivity of a host (receptor) for a particular target molecule or ion, the guest or ligand. An application is the development of chemical sensors. Finding a drug which either blocks a receptor, an antagonist which forms a strong complex the receptor, or activate it, an agonist, is an important pathway to drug discovery.

Complexes of Metals

Speciation diagram for aluminium in aqueous solution as a function of pH. A variety of hydroxo complexes are formed, including aluminium hydroxide, $(Al(OH)_3(s)$, which is insoluble at pH ~6.5.

The formation of a complex between a metal ion, M, and a ligand, L, is in fact usually a substitution reaction. For example, In aqueous solutions, metal ions will be present as aquo ions, so the reaction for the formation of the first complex could be written as:

$$[M(H_2O)_n] + L \rightleftharpoons [M(H_2O)_{n-1}L] + H_2O$$

However, since water is in vast excess, the concentration of water is usually assumed to be constant and is omitted from equilibrium constant expressions. Often, the metal and the ligand are in competition for protons. For the equilibrium,

$$p\,M + q\,L + r\,H \rightleftharpoons M_pL_qH_r$$

a stability constant can be defined as follows:

$$\beta_{pqr} = \frac{[M_pL_qH_r]}{[M]^p[L]^q[H]^r}$$

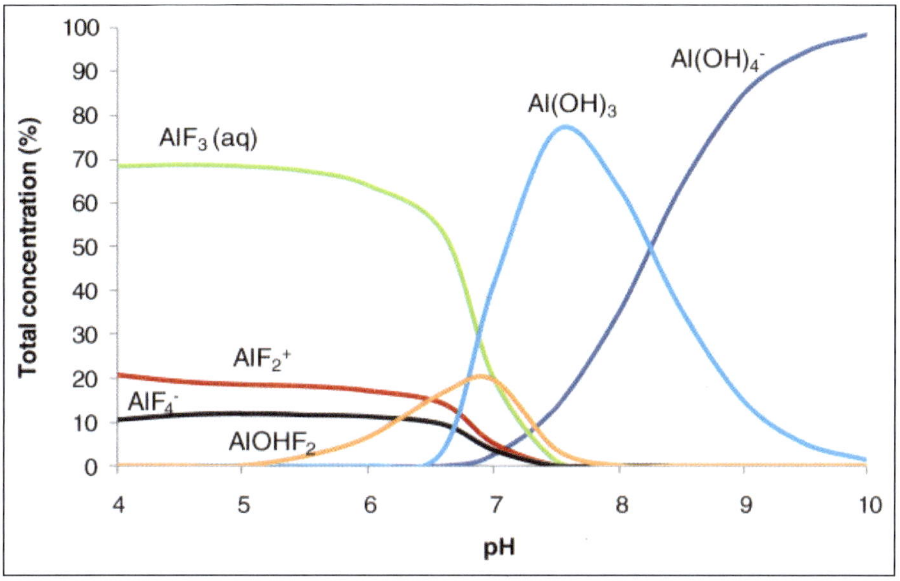

The definition can easily be extended to include any number of reagents. It includes hydroxide complexes because the concentration of the hydroxide ions is related to the concentration of hydrogen ions by the self-ionization of water:

$$[OH^-] = \frac{K_w}{[H^+]}$$

Stability constants defined in this way, are *association* constants. This can lead to some confusion as pK_a values are *dissociation* constants. In general purpose computer programs it is customary to define all constants as association constants. The relationship between the two types of constant is given in association and dissociation constants.

In biochemistry, an oxygen molecule can bind to an iron(II) atom in a heme prosthetic group in hemoglobin. The equilibrium is usually written, denoting hemoglobin by Hb, as,

$$Hb + O_2 \rightleftharpoons HbO_2$$

but this representation is incomplete as the Bohr effect shows that the equilibrium concentrations are pH-dependent. A better representation would be:

$$[HbH]^+ + O_2 \rightleftharpoons HbO_2 + H^+$$

as this shows that when hydrogen ion concentration increases the equilibrium is shifted to the left in accordance with Le Chatelier's principle. Hydrogen ion concentration can be increased by the presence of carbon dioxide, which behaves as a weak acid.

$$H_2O + CO_2 \rightleftharpoons HCO^-_3 + H^+$$

The iron atom can also bind to other molecules such as carbon monoxide. Cigarette smoke contains some carbon monoxide so the equilibrium,

$$HbO_2 + CO \rightleftharpoons Hb(CO) + O_2$$

is established in the blood of cigarette smokers.

Chelation therapy is based on the principle of using chelating ligands with a high binding selectivity for a particular metal to remove that metal from the human body.

Complexes with polyamino carboxylic acids find a wide range of applications. EDTA in particular is used extensively.

Redox Equilibria

A reduction–oxidation (redox) equilibrium can be handled in exactly the same way as any other chemical equilibrium. For example,

$$Fe^{2+} + Ce^{4+} \rightleftharpoons Fe^{3+} + Ce^{3+}; \; K = \frac{[Fe^{3+}][Ce^{3+}]}{[Fe^{2+}][Ce^{4+}]}$$

However, in the case of redox reactions it is convenient to split the overall reaction into two half-reactions. In this example:

$$Fe^{3+} + e^- \rightleftharpoons Fe^{2+}$$

$$Ce^{4+} + e^- \rightleftharpoons Ce^{3+}$$

The standard free energy change, which is related to the equilibrium constant by,

$$\Delta G^{\ominus} = -RT \ln K$$

can be split into two components,

$$\Delta G^{\ominus} = \Delta G^{\ominus}_{Fe} + \Delta G^{\ominus}_{Ce}$$

The concentration of free electrons is effectively zero as the electrons are transferred directly from the reductant to the oxidant. The standard electrode potential, E^o for the each half-reaction is related to the standard free energy change by,

$$\Delta G^{\ominus}_{Fe} = -nFE^0_{Fe}; \Delta G^{\ominus}_{Ce} = -nFE^0_{Ce}$$

where n is the number of electrons transferred and F is the Faraday constant. Now, the free energy for an actual reaction is given by,

$$\Delta G = \Delta G^{\ominus} + RT \ln Q$$

where R is the gas constant and Q a reaction quotient. Strictly speaking Q is a quotient of activities, but it is common practice to use concentrations instead of activities. Therefore:

$$E_{Fe} = E^0_{Fe} + \frac{RT}{nF} \ln \frac{[Fe^{3+}]}{[Fe^{2+}]}$$

For any half-reaction, the redox potential of an actual mixture is given by the generalized expression:

$$E = E^0 + \frac{RT}{nF}\ln\frac{[\text{oxidized species}]}{[\text{reduced species}]}$$

This is an example of the Nernst equation. The potential is known as a reduction potential. Standard electrode potentials are available in a table of values. Using these values, the actual electrode potential for a redox couple can be calculated as a function of the ratio of concentrations.

The equilibrium potential for a general redox half-reaction,

$$\alpha\,\mathrm{A} + \beta\,\mathrm{B}... + n\,\mathrm{e}^- \rightleftharpoons \sigma\,\mathrm{S} + \tau\,\mathrm{T}...$$

is given by,

$$E = E^{\ominus} + \frac{RT}{nF}\ln\frac{\{\mathrm{S}\}^{\sigma}\{\mathrm{T}\}^{\tau}...}{\{\mathrm{A}\}^{\alpha}\{\mathrm{B}\}^{\beta}...}$$

Use of this expression allows the effect of a species not involved in the redox reaction, such as the hydrogen ion in a half-reaction such as,

$$\mathrm{MnO^-_4} + 8\,\mathrm{H^+} + 5\,\mathrm{e^-} \rightleftharpoons \mathrm{Mn^{2+}} + 4\,\mathrm{H_2O}$$

to be taken into account.

The equilibrium constant for a full redox reaction can be obtained from the standard redox potentials of the constituent half-reactions. At equilibrium the potential for the two half-reactions must be equal to each other and, of course, the number of electrons exchanged must be the same in the two half reactions.

Redox equilibria play an important role in the electron transport chain. The various cytochromes in the chain have different standard redox potentials, each one adapted for a specific redox reaction. This allows, for example, atmospheric oxygen to be reduced in photosynthesis. A distinct family of cytochromes, the cytochrome P450 oxidases, are involved in steroidogenesis and detoxification.

Solubility

When a solute forms a saturated solution in a solvent, the concentration of the solute, at a given temperature, is determined by the equilibrium constant at that temperature.

$$\ln K = -RT\ln\left(\frac{\sum_k a_k^{m_k}(\text{solution})}{a(\text{solid})}\right)$$

The activity of a pure substance in the solid state is one, so the expression simplifies to:

$$\ln K = -RT\ln\left(\sum_k a_k^{m_k}(\text{solution})\right)$$

If the solute does not dissociate the summation is replaced by a single term, but if dissociation occurs, as with ionic substances:

$$K_{SP} = \prod_k a_k^{m_k}$$

For example, with Na_2SO_4, $m_1 = 2$ and $m_2 = 1$ so the solubility product is written as:

$$K_{SP} = [Na^+]^2[SO_4^{2-}]$$

Concentrations, indicated by, are usually used in place of activities, but activity must be taken into account of the presence of another salt with no ions in common, the so-called salt effect. When another salt is present that has an ion in common, the common-ion effect comes into play, reducing the solubility of the primary solute.

Partition

When a solution of a substance in one solvent is brought into equilibrium with a second solvent that is immiscible with the first solvent, the dissolved substance may be partitioned between the two solvents. The ratio of concentrations in the two solvents is known as a partition coefficient or distribution coefficient. The partition coefficient is defined as the ratio of the analytical concentrations of the solute in the two phases. By convention the value is reported in logarithmic form.

$$\log p = \log \frac{[\text{solute}]_{\text{organic phase}}}{[\text{solute}]_{\text{aqueous phase}}}$$

The partition coefficient is defined at a specified temperature and, if applicable, pH of the aqueous phase. Partition coefficients are very important in pharmacology because they determine the extent to which a substance can pass from the blood (an aqueous solution) through a cell wall which is like an organic solvent. They are usually measured using water and octanol as the two solvents. Many pharmaceutical compounds are weak acids or weak bases. Such a compound may exist with a different extent of protonation depending on pH and the acid dissociation constant. Because the organic phase has a low dielectric constant the species with no electrical charge will be the most likely one to pass from the aqueous phase to the organic phase. Even at pH 7–7.2, the range of biological pH values, the aqueous phase may support an equilibrium between more than one protonated form. log p is determined from the analytical concentration of the substance in the aqueous phase, that is, the sum of the concentration of the different species in equilibrium.

Solvent extraction is used extensively in separation and purification processes. In its simplest form a reaction is performed in an organic solvent and unwanted by-products are removed by extraction into water at a particular pH.

A metal ion may be extracted from an aqueous phase into an organic phase in which the salt is not soluble, by adding a ligand. The ligand, L^{a-}, forms a complex with the metal ion, M^{b+}, $[ML_x]^{(b-ax)+}$ which has a strongly hydrophobic outer surface. If the complex has no electrical charge it will be extracted relatively easily into the organic phase. If the complex is charged, it is extracted as an ion pair. The additional ligand is not always required. For example, uranyl nitrate, $UO_2(NO_3)_2$, is soluble in diethyl ether because the solvent itself acts as a ligand. This property was used in the past for separating uranium from other metals whose salts are not soluble in ether. Currently extraction into kerosene is preferred, using a ligand such as tri-*n*-butyl phohsphate, TBP. In the PUREX process, which is commonly used in nuclear reprocessing, uranium(VI) is extracted from strong nitric acid as the electrically neutral complex $[UO_2(TBP)_2(NO_3)_2]$. The strong nitric acid provides a high concentration of nitrate ions which pushes the equilibrium in favour of the weak nitrato complex. Uranium is recovered by back-extraction (stripping) into weak nitric acid. Plutonium(IV) forms a similar complex, $[PuO_2(TBP)_2(NO_3)_2]$ and the plutonium in this complex can be reduced to separate it from uranium.

Another important application of solvent extraction is in the separation of the lanthanoids. This process also uses TBP and the complexes are extracted into kerosene. Separation is achieved because the stability constant for the formation of the TBP complex increases as the size of the lanthanoid ion decreases.

An instance of ion-pair extraction is in the use of a ligand to enable oxidation by potassium permanganate, $KMnO_4$, in an organic solvent. $KMnO_4$ is not soluble in organic solvents. When a ligand, such as a crown ether is added to an aqueous solution of $KMnO_4$, it forms a hydrophobic complex with the potassium cation which allows the uncharged ion pair $[KL]^+[MnO_4]^-$ to be extracted into the organic solvent.

More complex partitioning problems (i.e. 3 or more phases present) can sometimes be handled with a fugacity capacity approach.

Chromatography

In chromatography substances are separated by partition between a stationary phase and a mobile phase. The analyte is dissolved in the mobile phase, and passes over the stationary phase. Separation occurs because of differing affinities of the analytes for the stationary phase. A distribution constant, K_d can be defined as,

$$K_d = \frac{a_s}{a_m}$$

where a_s and a_m are the equilibrium activities in the stationary and mobile phases respectively. It can be shown that the rate of migration, v, is related to the distribution constant by:

$$\bar{v} \propto \frac{1}{1 + fK_d}.$$

Where, f is a factor which depends on the volumes of the two phases. Thus, the higher the affinity of the solute for the stationary phase, the slower the migration rate.

There is a wide variety of chromatographic techniques, depending on the nature of the stationary and mobile phases. When the stationary phase is solid, the analyte may form a complex with it. A water softener functions by selective complexation with a sulfonate ion exchange resin. Sodium ions form relatively weak complexes with the resin. When hard water is passed through the resin, the divalent ions of magnesium and calcium displace the sodium ions and are retained on the resin, R.

$$RNa + M^{2+} \rightleftharpoons RM^+ + Na^+$$

The water coming out of the column is relatively rich in sodium ions and poor in calcium and magnesium which are retained on the column. The column is regenerated by passing a strong solution of sodium chloride through it, so that the resin–sodium complex is again formed on the column. Ion-exchange chromatography utilizes a resin such as chelex 100 in which iminodiacetate residues, attached to a polymer backbone, form chelate complexes of differing strengths with different metal ions, allowing the ions such as Cu^{2+} and Ni^{2+} to be separated chromatographically.

Another example of complex formation is in chiral chromatography in which is used to separate enantiomers from each other. The stationary phase is itself chiral and forms complexes selectively with the enantiomers. In other types of chromatography with a solid stationary phase, such as thin layer chromatography the analyte is selectively adsorbed onto the solid.

In gas–liquid chromatography (GLC) the stationary phase is a liquid such as polydimethylsiloxane, coated on a glass tube. Separation is achieved because the various components in the gas have different solubility in the stationary phase. GLC can be used to separate literally hundreds of components in a gas mixture such as cigarette smoke or essential oils, such as lavender oil.

Law of Chemical Equilibrium

Representation of the attainment of chemical equilibrium:

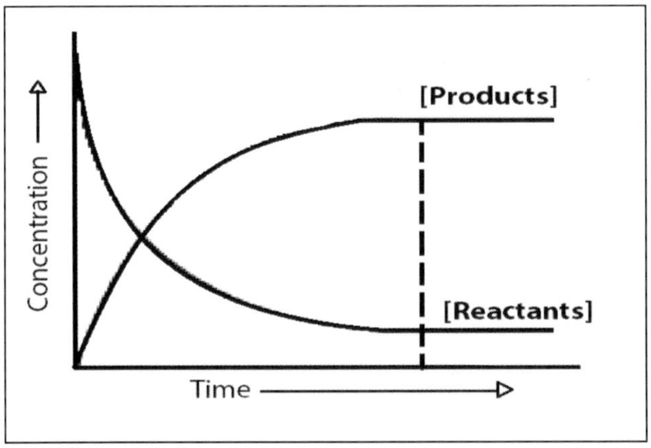

The equilibrium constant is defined as the product of the molar concentration of the products which is each raised to the power equal to its stoichiometric coefficients divided by the product of the molar concentrations of the reactants, each raised to the power equal to its stoichiometric coefficients is constant at constant temperature. This equilibrium constant can be simply expressed in terms of the partial pressures of the reactants and the products. However, if it is expressed in terms of the partial pressure, it is denoted by K_p.

Law of mass action also forms the basis which states that the rate of a chemical reaction is directly proportional to the product of the concentrations of the reactants raised to their respective stoichiometric coefficients. Therefore, given the reaction:

$$aA(g) + bB(g) \Leftrightarrow cC(g) + dD(g)$$

By using the law of mass action here,

- The forward reaction rate would be $k^+ [A]^a[B]^b$.
- The backward reaction rate would be $k^+ [C]^c[D]^d$.

Where, [A], [B], [C] and [D] being the active masses and k_+ and k_- are rate constants of forward and backward reactions, also the a, b, c, d are the stoichiometric coefficients related to A, B, C and D respectively. However, at the equilibrium – the forward and the backward rates are equal, stating –

Rate of forward reaction = Rate of backward reaction,

$$K_f [A]^a [B]^b = K_b [C]^c [D]^d$$

$$\frac{K_f}{K_b} = \frac{[C]^c [D]^d}{[A]^a [B]^b}$$

or

$$K_c = \frac{[C]^c [D]^d}{[A]^a [B]^b}$$

or

$$K_c = \frac{K_f}{K_b}$$

where,

K_c is the equilibrium constant expressed in terms of the molar concentrations. The equation $K_c = [C]^c \cdot [D]^d/[A]^a \cdot [B]^b$ or $K_c = K_f/K_b$ is the Law of Chemical Equilibrium. The equilibrium constant is therefore related to the standard Gibbs free energy change for the reaction which is stated by the equation,

$$\S G^o = -RT \ln K_{eq}$$

where T states the temperature, R is the universal gas constant and K_{eq} is the equilibrium constant.

The Equilibrium Constant,

$$N_2O_4(g) \rightleftharpoons 2NO_2(g)$$

Forward reaction:

$$N_2O_4 (g) \rightarrow 2\, NO_2 (g).$$

Rate Law = $K_f [N_2O_4]$.

Reverse reaction:

$$N_2O_4\,(g) \leftarrow 2\,NO_2\,(g).$$

Rate Law = $K_f\,[N_2]^2$.

Equilibrium Constant

The equilibrium constant of a chemical reaction is the value of its reaction quotient at chemical equilibrium, a state approached by a dynamic chemical system after sufficient time has elapsed at which its composition has no measurable tendency towards further change. For a given set of reaction conditions, the equilibrium constant is independent of the initial analytical concentrations of the reactant and product species in the mixture. Thus, given the initial composition of a system, known equilibrium constant values can be used to determine the composition of the system at equilibrium. However, reaction parameters like temperature, solvent, and ionic strength may all influence the value of the equilibrium constant.

A knowledge of equilibrium constants is essential for the understanding of many chemical systems, as well as biochemical processes such as oxygen transport by hemoglobin in blood and acid-base homeostasis in the human body.

Stability constants, formation constants, binding constants, association constants and dissociation constants are all types of equilibrium constants.

Basic Properties

For a system undergoing a reversible reaction described by the general chemical equation,

$$\alpha \mathrm{A} + \beta \mathrm{B} + \cdots \rightleftharpoons \rho \mathrm{R} + \sigma \mathrm{S} + \cdots$$

a thermodynamic equilibrium constant, denoted by $K^{\ominus}$, is defined to be the value of the reaction quotient Q_t when forward and reverse reactions occur at the same rate. At chemical equilibrium, the chemical composition of the mixture does not change with time and the Gibbs free energy change ΔG for the reaction is zero. If the composition of a mixture at equilibrium is changed by addition of some reagent, a new equilibrium position will be reached, given enough time. An equilibrium constant is related to the composition of the mixture at equilibrium by,

$$K^{\ominus} = \frac{\{\mathrm{R}\}^{\rho}\{\mathrm{S}\}^{\sigma}...}{\{\mathrm{A}\}^{\alpha}\{\mathrm{B}\}^{\beta}...} = \frac{[\mathrm{R}]^{\rho}[\mathrm{S}]^{\sigma}...}{[\mathrm{A}]^{\alpha}[\mathrm{B}]^{\beta}...} \times \Gamma,$$

$$\Gamma = \frac{\gamma_R^{\rho}\gamma_S^{\sigma}\cdots}{\gamma_A^{\alpha}\gamma_B^{\beta}\cdots},$$

where {X} denotes the thermodynamic activity of reagent X at equilibrium, [X] the corresponding concentration, and γ the corresponding activity coefficient. If it can be assumed that the quotient of activity coefficients, Γ, is constant over a range of experimental conditions, such as pH, then an equilibrium constant can be derived as a quotient of concentrations.

$$K_c = K^{\ominus} / \Gamma = \frac{[R]^{\rho}[S]^{\sigma}...}{[A]^{\alpha}[B]^{\beta}...}$$

An equilibrium constant is related to the standard Gibbs free energy change of reaction $\Delta G^{\ominus}$ by,

$$\Delta G^{\ominus} = -RT \ln K^{\ominus}$$

where R is the universal gas constant, T is the absolute temperature (in kelvins), and *ln* is a natural logarithm. This expression implies that $K^{\ominus}$ must be a pure number and cannot have a dimension, since logarithms can only be taken of pure numbers. K_c must also be a pure number. On the other hand, the reaction quotient at equilibrium,

$$\frac{[R]^{\rho}[S]^{\sigma}...}{[A]^{\alpha}[B]^{\beta}...}$$

The above equation does have the dimension of concentration raised to some power. Such reaction quotients are often referred to, in the biochemical literature, as equilibrium constants.

For an equilibrium mixture of gases, an equilibrium constant can be defined in terms of partial pressure or fugacity.

An equilibrium constant is related to the forward and backward rate constants, k_f and k_r of the reactions involved in reaching equilibrium:

$$K^{\ominus} = \frac{k_f}{k_r}.$$

Types of Equilibrium Constants

Cumulative and Stepwise Formation Constants

A cumulative or overall constant, given the symbol β, is the constant for the formation of a complex from reagents. For example, the cumulative constant for the formation of ML_2 is given by:

$$M + 2\,L \rightleftharpoons ML_2;\ [ML_2] = \beta_{12}[M][L]^2$$

The stepwise constant, K, for the formation of the same complex from ML and L is given by:

$$ML + L \rightleftharpoons ML_2;\ [ML_2] = K[ML][L] = K\beta_{11}[M][L]^2$$

It follows that,

$$\beta_{12} = K\beta_{11}$$

A cumulative constant can always be expressed as the product of stepwise constants. There is no agreed notation for stepwise constants, though a symbol such as K^{L}_{ML} is sometimes found in the literature. It is best always to define each stability constant by reference to an equilibrium expression.

Competition Method

A particular use of a stepwise constant is in the determination of stability constant values outside the normal range for a given method. For example, EDTA complexes of many metals are outside the range for the potentiometric method. The stability constants for those complexes were determined by competition with a weaker ligand.

$$\mathrm{ML}+\mathrm{L}'\rightleftharpoons \mathrm{ML}'+\mathrm{L}\;[\mathrm{ML}']=K\frac{[\mathrm{ML}][\mathrm{L}']}{[\mathrm{L}]}=K\frac{\beta_{\mathrm{ML}}[\mathrm{M}][\mathrm{L}][\mathrm{L}']}{[\mathrm{L}]}=K\beta_{\mathrm{ML}}[\mathrm{M}][\mathrm{L}'];\quad \beta_{\mathrm{ML}'}=K\beta_{\mathrm{ML}}$$

The formation constant of $[Pd(CN)_4]^{2-}$ was determined by the competition method.

Association and Dissociation Constants

In organic chemistry and biochemistry it is customary to use pK_a values for acid dissociation equilibria,

$$pK_a=-\log K_{\mathrm{diss}}=\log\left(\frac{1}{K_{\mathrm{diss}}}\right)$$

where log denotes a logarithm to base 10 or common logarithm, and K_{diss} is a stepwise acid dissociation constant. For bases, the base association constant, pK_b is used. For any given acid or base the two constants are related by $pK_a + pK_b = pK_w$, so pK_a can always be used in calculations.

On the other hand, stability constants for metal complexes, and binding constants for host–guest complexes are generally expressed as association constants. When considering equilibria such as,

$$\mathrm{M} + \mathrm{HL} \rightleftharpoons \mathrm{ML} + \mathrm{H}$$

it is customary to use association constants for both ML and HL. Also, in generalized computer programs dealing with equilibrium constants it is general practice to use cumulative constants rather than stepwise constants and to omit ionic charges from equilibrium expressions. For example, if NTA, nitrilotriacetic acid, $N(CH_2CO_2H)_3$ is designated as H_3L and forms complexes ML and MHL with a metal ion M, the following expressions would apply for the dissociation constants.

$$\mathrm{H_3L}\rightleftharpoons \mathrm{H_2L}+\mathrm{H};\quad pK_1=-\log\left(\frac{[\mathrm{H_2L}][\mathrm{H}]}{[\mathrm{H_3L}]}\right)$$

$$\mathrm{H_2L}\rightleftharpoons \mathrm{HL}+\mathrm{H};\quad pK_2=-\log\left(\frac{[\mathrm{HL}][\mathrm{H}]}{[\mathrm{H_2L}]}\right)$$

$$\mathrm{HL}\rightleftharpoons \mathrm{L}+\mathrm{H};\quad pK_3=-\log\left(\frac{[\mathrm{L}][\mathrm{H}]}{[\mathrm{HL}]}\right)$$

The cumulative association constants can be expressed as:

$$\mathrm{L}+\mathrm{H}\rightleftharpoons \mathrm{HL};\quad \log\beta_{011}=\log\left(\frac{[\mathrm{HL}]}{[\mathrm{L}][\mathrm{H}]}\right)=pK_3$$

$$L + 2H \rightleftharpoons H_2L; \qquad \log \beta_{012} = \log\left(\frac{[H_2L]}{[L][H]^2}\right) = pK_3 + pK_2$$

$$L + 3H \rightleftharpoons H_3L; \qquad \log \beta_{013} = \log\left(\frac{[H_3L]}{[L][H]^3}\right) = pK_3 + pK_2 + pK_1$$

$$M + L \rightleftharpoons ML; \qquad \log \beta_{110} = \log\left(\frac{[ML]}{[M][L]}\right)$$

$$M + L + H \rightleftharpoons MLH; \quad \log \beta_{111} = \log\left(\frac{[MLH]}{[M][L][H]}\right)$$

Micro-constants

When two or more sites in an asymmetrical molecule may be involved in an equilibrium reaction there are more than one possible equilibrium constants. For example, the molecule L-DOPA has two non-equivalent hydroxyl groups which may be deprotonated. Denoting L-DOPA as LH_2, the following diagram shows all the species that may be formed ($X = CH_2CH(NH_2)CO_2H$):

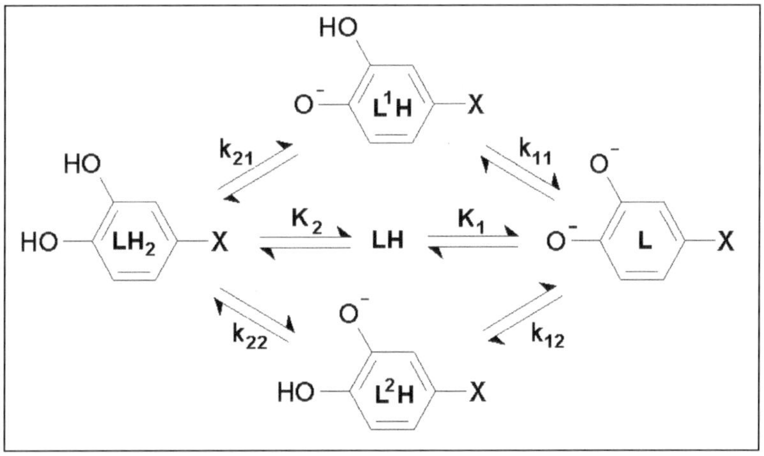

The first protonation constants are:

$[L^1H] = k_{11}[L][H]$, $[L^2H] = k_{12}[L][H]$

The concentration of LH^- is the sum of the concentrations of the two micro-species. Therefore, the equilibrium constant for the reaction, the macro-constant, is the sum of the micro-constants.

$$K_1 = k_{11} + k_{12}$$

In the same way,

$$K_2 = k_{21} + k_{22}$$

Lastly, the cumulative constant is:

$$\beta_2 = K_1K_2 = k_{11}k_{21} = k_{12}k_{22}$$

Thus, although there are seven micro-and macro-constants, only three of them are mutually independent. Moreover, the isomerization constant, K_i, is equal to the ratio of the microconstants.

$$K_i = k_{11}/k_{12}$$

In L-DOPA the isomerization constant is 0.9, so the micro-species L^1H and L^2H have almost equal concentrations at all pH values.

In general a macro-constant is equal to the sum of all the micro-constants and the occupancy of each site is proportional to the micro-constant. The site of protonation can be very important, for example, for biological activity.

Micro-constants cannot be determined individually by the usual methods, which give macro-constants. Methods which have been used to determine micro-constants include:

- Blocking one of the sites, for example by methylation of a hydroxyl group, to determine one of the micro-constants.
- Using a spectroscopic technique, such as infrared spectroscopy, where the different micro-species give different signals.
- Applying mathematical procedures to ^{13}C NMR data.

pH Considerations: Bronsted Constants

pH is defined in terms of the activity of the hydrogen ion:

$$\mathrm{pH} = -\log_{10}\{H^+\}$$

In the approximation of ideal behavior, activity is replaced by concentration. pH is measured by means of a glass electrode, a mixed equilibrium constant, also known as a Bronsted constant, may result.

$$\mathrm{HL} \rightleftharpoons \mathrm{L} + \mathrm{H};\ \mathrm{p}K = -\log\left(\frac{[\mathrm{L}]\{\mathrm{H}\}}{[\mathrm{HL}]}\right)$$

It all depends on whether the electrode is calibrated by reference to solutions of known activity or known concentration. In the latter case the equilibrium constant would be a concentration quotient. If the electrode is calibrated in terms of known hydrogen ion concentrations it would be better to write p[H] rather than pH, but this suggestion is not generally adopted.

Hydrolysis Constants

In aqueous solution the concentration of the hydroxide ion is related to the concentration of the hydrogen ion by:

$$K_W = [\mathrm{H}][\mathrm{OH}]$$

$$[\mathrm{OH}] = K_W[\mathrm{H}]^{-1}$$

The first step in metal ion hydrolysis can be expressed in two different ways:

$$\begin{cases} M(H_2O) \rightleftharpoons M(OH) + H; & [M(OH)] = \beta^*[M][H]^{-1} \\ M + OH \rightleftharpoons M(OH); & [M(OH)] = K[M][OH] = KK_W[M][H]^{-1} \end{cases}$$

It follows that $\beta^* = KK_W$. Hydrolysis constants are usually reported in the β^* form and therefore often have values much less than 1. For example, if log K = 4 and log K_W = −14, log β^* = 4 + (−14) = −10 so that $\beta^* = 10^{-10}$. In general when the hydrolysis product contains n hydroxide groups log β^* = log K + n log K_W.

Conditional Constants

Conditional constants, also known as apparent constants, are concentration quotients which are not true equilibrium constants but can be derived from them. A very common instance is where pH is fixed at a particular value. For example, in the case of iron(III) interacting with EDTA, a conditional constant could be defined by:

$$K_{cond} = \frac{[\text{Total Fe bound to EDTA}]}{[\text{Total Fe not bound to EDTA}] \times [\text{Total EDTA not bound to Fe}]}$$

This conditional constant will vary with pH. It has a maximum at a certain pH. That is the pH where the ligand sequesters the metal most effectively.

In biochemistry equilibrium constants are often measured at a pH fixed by means of a buffer solution. Such constants are, conditional and different values may be obtained when using different buffers.

Gas-phase Equilibria

For equilibria in a gas phase, fugacity, f, is used in place of activity. However, fugacity has the dimension of pressure, so it must be divided by a standard pressure, usually 1 bar, in order to produce a dimensionless quantity, f/p^o. An equilibrium constant is expressed in terms of the dimensionless quantity. For example, for the equilibrium $2NO_2 \rightleftharpoons N_2O_4$,

$$\frac{f_{N_2O_4}}{p^\ominus} = K\left(\frac{f_{NO_2}}{p^\ominus}\right)^2$$

Fugacity is related to partial pressure, p, by a dimensionless fugacity coefficient ϕ: $f = \phi p$. Thus, for the example,

$$K = \frac{\phi_{N_2O_4} p_{N_2O_4} / p^\ominus}{\left(\phi_{NO_2} p_{NO_2} / p^\ominus\right)^2}$$

Usually the standard pressure is omitted from such expressions. Expressions for equilibrium constants in the gas phase then resemble the expression for solution equilibria with fugacity coefficient in place of activity coefficient and partial pressure in place of concentration.

$$K = \frac{\phi_{N_2O_4} p_{N_2O_4}}{\left(\phi_{NO_2} p_{NO_2}\right)^2}$$

Thermodynamic basis for Equilibrium Constant Expressions

Thermodynamic equilibrium is characterized by the free energy for the whole (closed) system being a minimum. For systems at constant temperature and pressure the Gibbs free energy is minimum. The slope of the reaction free energy with respect to the reaction coordinate, ξ, is zero when the free energy is at its minimum value.

$$\left(\frac{\partial G}{\partial \xi}\right)_{T,P} = 0$$

The free energy change, δG_r, can be expressed as a weighted sum of chemical potentials, the partial molar free energies of the species in equilibrium. The potential, μ_i, of the *i*th species in a chemical reaction is the partial derivative of the free energy with respect to the number of moles of that species, N_i:

$$\mu_i = \left(\frac{\partial G}{\partial N_i}\right)_{T,P}$$

A general chemical equilibrium can be written as,

$$\sum_j n_j \text{Reactant}_j \rightleftharpoons \sum_k m_k \text{Product}_k$$

where n_j are the stoichiometric coefficients of the reactants in the equilibrium equation, and m_j are the coefficients of the products. At equilibrium,

$$\sum_k m_k \mu_k = \sum_j n_j \mu_j$$

The chemical potential, μ_i, of the *i*th species can be calculated in terms of its activity, a_i.

$$\mu_i = \mu_i^{\ominus} + RT \ln a_i$$

where $\mu_i^{\ominus}$ is the standard chemical potential of the species, R is the gas constant and T is the temperature. Setting the sum for the reactants *j* to be equal to the sum for the products, *k*, so that $\delta G_r(\text{Eq}) = 0$:

$$\sum_j n_j (\mu_j^{\ominus} + RT \ln a_j) = \sum_k m_k (\mu_k^{\ominus} + RT \ln a_k)$$

Rearranging the terms,

$$\sum_k m_k \mu_k^\ominus - \sum_j n_j \mu_j^\ominus = -RT\left(\sum_k \ln a_k^{\ m_k} - \sum_j \ln a_j^{\ n_j}\right)$$

$$\Delta G^\ominus = -RT \ln K.$$

This relates the standard Gibbs free energy change, ΔG^o to an equilibrium constant, K, the reaction quotient of activity values at equilibrium.

$$\Delta G^\ominus = \sum_k m_k \mu_k^\ominus - \sum_j n_j \mu_j^\ominus$$

$$\ln K = \sum_k \ln a_k^{\ m_k} - \sum_j \ln a_j^{\ n_j}; K = \frac{\prod_k a_k^{\ m_k}}{\prod_j a_j^{\ n_j}} \equiv \frac{\{R\}^\rho \{S\}^\sigma ...}{\{A\}^\alpha \{B\}^\beta ...}$$

Equivalence of Thermodynamic and Kinetic Expressions for Equilibrium Constants

At equilibrium the rate of the forward reaction is equal to the backward reaction rate. A simple reaction, such as ester hydrolysis,

$$\mathrm{AB} + \mathrm{H_2O} \rightleftharpoons \mathrm{AH} + \mathrm{B(OH)}$$

It has reaction rates given by expressions:

$$\text{forward rate} = k_f[\mathrm{AB}][\mathrm{H_2O}]$$

$$\text{backward rate} = k_b[\mathrm{AH}][\mathrm{B(OH)}]$$

According to Guldberg and Waage, equilibrium is attained when the forward and backward reaction rates are equal to each other. In these circumstances, an equilibrium constant is defined to be equal to the ratio of the forward and backward reaction rate constants.

$$K = \frac{k_f}{k_b} = \frac{[\mathrm{AH}][\mathrm{B(OH)}]}{[\mathrm{AB}][\mathrm{H_2O}]}$$

For convenience the concentration of water may be taken to be constant, resulting in the simpler expression:

$$K^c = \frac{[\mathrm{AH}][\mathrm{B(OH)}]}{[\mathrm{AB}]}$$

This particular concentration quotient, K^c, has the dimension of concentration, but the thermodynamic equilibrium constant, K, is always dimensionless.

Unknown Activity Coefficient Values

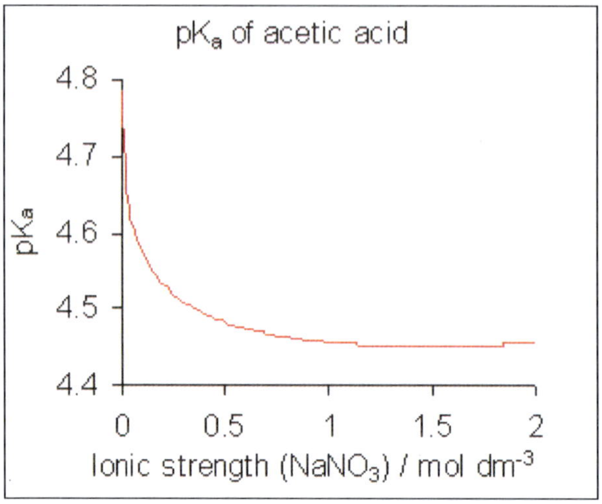

Variation of log *Kc* of acetic acid with ionic strength.

It is very rare for activity coefficient values to have been determined experimentally for a system at equilibrium. There are three options for dealing with the situation where activity coefficient values are not known from experimental measurements.

1. Use calculated activity coefficients, together with concentrations of reactants. For equilibria in solution estimates of the activity coefficients of charged species can be obtained using Debye–Hückel theory, an extended version, or SIT theory. For uncharged species, the activity coefficient γ_o mostly follows a "salting-out" model: $\log_{10} \gamma_o = bI$ where I stands for ionic strength.

2. Assume that the activity coefficients are all equal to 1. This is acceptable when all concentrations are very low.

3. For equilibria in solution use a medium of high ionic strength. In effect this redefines the standard state as referring to the medium. Activity coefficients in the standard state are, equal to 1. The value of an equilibrium constant determined in this manner is dependent on the ionic strength. When published constants refer to an ionic strength other than the one required for a particular application, they may be adjusted by means of specific ion theory (SIT) and other theories.

Dimensionality

An equilibrium constant is related to the standard Gibbs free energy change, $\Delta G^{\ominus}$, for the reaction by the expression:

$$\Delta G^{\ominus} = -RT \ln K$$

Therefore K, must be a number from which a logarithm can be derived. In the case of a simple equilibrium,

$$\mathrm{A + B \rightleftharpoons AB}$$

the thermodynamic equilibrium constant is defined in terms of the activities, {AB}, {A} and {B}, of the species in equilibrium with each other.

$$K = \frac{\{AB\}}{\{A\}\{B\}}$$

Now, each activity term can be expressed as a product of a concentration $[X]$ and a corresponding activity coefficient, $\gamma(X)$. Therefore,

$$K = \frac{[AB]}{[A][B]} \times \frac{\gamma(AB)}{\gamma(A)\gamma(B)} = \frac{[AB]}{[A][B]} \times \Gamma$$

When the quotient of activity coefficients, Γ is set equal to 1:

$$K = \frac{[AB]}{[A][B]}$$

K then appears to have the dimension of 1/concentration. This is what usually happens in practice when an equilibrium constant is calculated as a quotient of concentration values. The assumption underlying this practice is that the quotient of activities is constant under the conditions in which the equilibrium constant value is determined. There conditions are usually achieved by keeping the reaction temperature constant and by using a medium of relatively high ionic strength as the solvent. It is not unusual, particularly in texts relating to biochemical equilibria, to see a value quoted with a dimension. The justification for this practice is that the concentration scale used may be either mol dm^{-3} or mmol dm^{-3}, so that the concentration unit has to be stated in order to avoid there being any ambiguity.

When the concentration values are measured on the mole fraction scale all concentrations and activity coefficients are dimensionless quantities.

In general equilibria between two reagents can be expressed as,

$$p\,\mathrm{A} + q\,\mathrm{B} \rightleftharpoons \mathrm{A}_p\,\mathrm{B}_q$$

in which case the equilibrium constant is defined, in terms of numerical concentration values, as:

$$K = \frac{[\mathrm{A_pB_q}]}{[\mathrm{A}]^{\mathrm{p}}[\mathrm{B}]^{\mathrm{q}}}$$

The apparent dimension of this K value is $\text{concentration}^{1-p-q}$; this may be written as $M^{(1-p-q)}$ or $mM^{(1-p-q)}$, where the symbol M signifies a molar concentration (1M = 1 mol dm^{-3}). The apparent dimension of a dissociation constant is the reciprocal of the apparent dimension of the corresponding association constant, and vice versa.

When discussing the thermodynamics of chemical equilibria it is necessary to take dimensionality into account. There are two possible approaches.

1. Set the dimension of Γ to be the reciprocal of the dimension of the concentration quotient. This is almost universal practice in the field of stability constant determinations. The

"equilibrium constant" $\frac{K}{\Gamma}$, is dimensionless. It will be a function of the ionic strength of the medium used for the determination. Setting the numerical value of Γ to be 1 is equivalent to re-defining the standard states.

2. Replace each concentration term $[X]$ by the dimensionless quotient $\frac{[X]}{[X^0]}$, where $[X^0]$ is the concentration of reagent X in its standard state. The numerical value of $\gamma(X^0)$ is 1, so Γ also has a numerical value of 1.

In both approaches the numerical value of the stability constant is unchanged. The first is more useful for practical purposes; in fact, the unit of the concentration quotient is often attached to a published stability constant value in the biochemical literature. The second approach is consistent with the standard exposition of Debye–Hückel theory, where $\gamma(AB)$, *etc.* are taken to be pure numbers.

Water as both Reactant and Solvent

For reactions in aqueous solution, such as an acid dissociation reaction,

$$AH + H_2O \rightleftharpoons A^- + H_3O^+$$

the concentration of water may be taken as being constant and the formation of the hydronium ion is implicit.

$$AH \rightleftharpoons A^- + H^+$$

Water concentration is omitted from expressions defining equilibrium constants, except when solutions are very concentrated.

$$K = \frac{[A][H]}{[AH]} \text{ (K defined as a dissociation constant)}$$

Similar considerations apply to metal ion hydrolysis reactions.

Enthalpy and Entropy: Temperature Dependence

If both the equilibrium constant, K and the standard enthalpy change, $\Delta H^\ominus$, for a reaction have been determined experimentally, the standard entropy change for the reaction is easily derived. Since $\Delta G = \Delta H - T\Delta S$ and $\Delta G = -RT \ln K$:

$$\Delta S^\ominus = \frac{\Delta H^\ominus + RT \ln K}{T}$$

To a first approximation the standard enthalpy change is independent of temperature. Using this approximation, definite integration of the Van 't Hoff equation,

$$\Delta H^\ominus = -R \frac{d \ln K}{d(1/T)}$$

gives,

$$\ln K_2 = \ln K_1 - \frac{\Delta H^{\ominus}}{R}\left(\frac{1}{T_2} - \frac{1}{T_1}\right)$$

This equation can be used to calculate the value of log K at a temperature, T_2, knowing the value at temperature T_1.

The Van 't Hoff equation also shows that, for an exothermic reaction ($\Delta H < 0$), when temperature increases *K* decreases and when temperature decreases *K* increases, in accordance with Le Chatelier's principle. The reverse applies when the reaction is endothermic.

When *K* has been determined at more than two temperatures, a straight line fitting procedure may be applied to a plot of $\ln K$ against $1/T$ to obtain a value for $\Delta H^{\ominus}$. Error propagation theory can be used to show that, with this procedure, the error on the calculated $\Delta H^{\ominus}$ value is much greater than the error on individual log K values. Consequently, K needs to be determined to high precision when using this method.For example, with a silver ion-selective electrode each log K value was determined with a precision of ca. 0.001 and the method was applied successfully.

Standard thermodynamic arguments can be used to show that, more generally, enthalpy will change with temperature.

$$\left(\frac{\partial H}{\partial T}\right)_p = C_p$$

where C_p is the heat capacity at constant pressure.

A more Complex Formulation

The calculation of *K* at a particular temperature from a known *K* at another given temperature can be approached as follows if standard thermodynamic properties are available. The effect of temperature on equilibrium constant is equivalent to the effect of temperature on Gibbs energy because:

$$\ln K = \frac{-\Delta_r G^{\ominus}}{RT}$$

where $\Delta_r G^{\ominus}$ is the reaction standard Gibbs energy, which is the sum of the standard Gibbs energies of the reaction products minus the sum of standard Gibbs energies of reactants.

Here, the term "standard" denotes the ideal behavior (i.e., an infinite dilution) and a hypothetical standard concentration (typically 1 mol/kg). It does not imply any particular temperature or pressure because, although contrary to IUPAC recommendation, it is more convenient when describing aqueous systems over wide temperature and pressure ranges.

The standard Gibbs energy (for each species or for the entire reaction) can be represented as:

$$G_{T_2}^{\ominus} = G_{T_1}^{\ominus} - S_{T_1}^{\ominus}(T_2 - T_1) - T_2\int_{T_1}^{T_2} \frac{C_p^{\ominus}}{T} dT + \int_{T_1}^{T_2} C_p^{\ominus}\, dT$$

In the above equation, the effect of temperature on Gibbs energy (and thus on the equilibrium constant) is ascribed entirely to heat capacity. To evaluate the integrals in this equation, the form of the dependence of heat capacity on temperature needs to be known.

If the standard molar heat capacity $C_p^{\ominus}$ can be approximated by some analytic function of temperature (e.g. the Shomate equation), then the integrals involved in calculating other parameters may be solved to yield analytic expressions for them. For example, using approximations of the following forms:

- For pure substances (solids, gas, liquid):

  $$C_p^{\ominus} \approx A + BT + CT^{-2}$$

- For ionic species at $T < 200$ °C:

  $$C_p^{\ominus} \approx (4.186a + b\breve{S}_{T_1}^{\ominus})\frac{(T_2 - T_1)}{\ln\left(\frac{T_2}{T_1}\right)}$$

Then the integrals can be evaluated and the following final form is obtained:

$$G_{T_2}^{\ominus} \approx G_{T_1}^{\ominus} + (C_p^{\ominus} - S_{T_1}^{\ominus})(T_2 - T_1) - T_2 \ln\left(\frac{T_2}{T_1}\right)C_p^{\ominus}$$

The constants A, B, C, a, b and the absolute entropy, $\breve{S}_{298K}^{\ominus}$, required for evaluation of $C_p^{\ominus}(T)$, as well as the values of $G_{298\text{ K}}$ and $S_{298\text{ K}}$ for many species are tabulated in the literature.

Pressure Dependence

The pressure dependence of the equilibrium constant is usually weak in the range of pressures normally encountered in industry, and therefore, it is usually neglected in practice. This is true for condensed reactant/products (i.e., when reactants and products are solids or liquid) as well as gaseous ones.

For a gaseous-reaction example, one may consider the well-studied reaction of hydrogen with nitrogen to produce ammonia:

$$N_2 + 3\,H_2 \rightleftharpoons 2\,NH_3$$

If the pressure is increased by the addition of an inert gas, then neither the composition at equilibrium nor the equilibrium constant are appreciably affected (because the partial pressures remain constant, assuming an ideal-gas behavior of all gases involved). However, the composition at equilibrium will depend appreciably on pressure when:

- The pressure is changed by compression or expansion of the gaseous reacting system,
- The reaction results in the change of the number of moles of gas in the system.

In the above reaction, the number of moles changes from 4 to 2, and an increase of pressure by system compression will result in appreciably more ammonia in the equilibrium mixture. In the general case of a gaseous reaction,

$$\alpha\,\mathrm{A} + \beta\,\mathrm{B} \rightleftharpoons \sigma\,\mathrm{S} + \tau\,\mathrm{T}$$

the change of mixture composition with pressure can be quantified using,

$$K_p = \frac{p_\mathrm{S}^{\ \sigma} p_\mathrm{T}^{\ \tau}}{p_\mathrm{A}^{\ \alpha} p_\mathrm{B}^{\ \beta}} = \frac{X_\mathrm{S}^{\ \sigma} X_\mathrm{T}^{\ \tau}}{X_\mathrm{A}^{\ \alpha} X_\mathrm{B}^{\ \beta}} P^{\sigma+\tau-\alpha-\beta} = K_X P^{\sigma+\tau-\alpha-\beta}$$

where p denote the partial pressures and X the mole fractions of the components, P is the total system pressure, K_p is the equilibrium constant expressed in terms of partial pressures and K_X is the equilibrium constant expressed in terms of mole fractions.

The change in composition is in accordance with Le Chatelier's principle and does not involve any change of the equilibrium constant with the total system pressure. Indeed, for ideal-gas reactions K_p is independent of pressure.

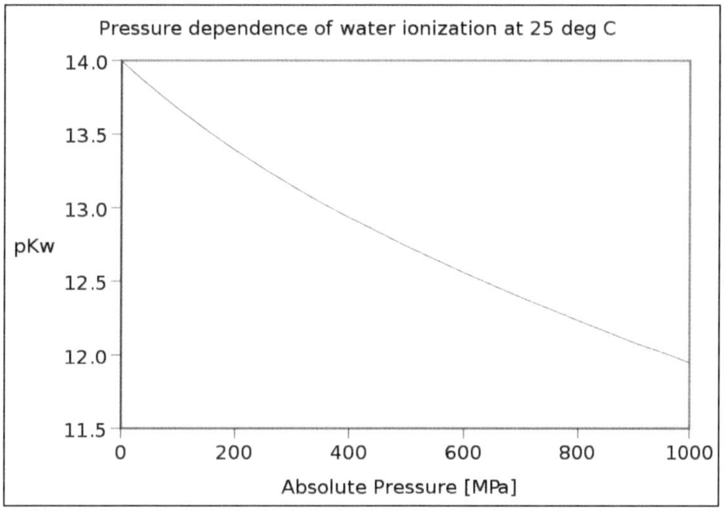

Pressure dependence of the water ionization constant at 25 °C. In general, ionization in aqueous solutions tends to increase with increasing pressure.

In a condensed phase, the pressure dependence of the equilibrium constant is associated with the reaction volume. For reaction,

$$\alpha\,\mathrm{A} + \beta\,\mathrm{B} \rightleftharpoons \sigma\,\mathrm{S} + \tau\,\mathrm{T}$$

the reaction volume is,

$$K_p = \frac{p_\mathrm{S}^{\ \sigma} p_\mathrm{T}^{\ \tau}}{p_\mathrm{A}^{\ \alpha} p_\mathrm{B}^{\ \beta}} = \frac{X_\mathrm{S}^{\ \sigma} X_\mathrm{T}^{\ \tau}}{X_\mathrm{A}^{\ \alpha} X_\mathrm{B}^{\ \beta}} P^{\sigma+\tau-\alpha-\beta} = K_X P^{\sigma+\tau-\alpha-\beta}$$

where $\overline{V}$ denotes a partial molar volume of a reactant or a product.

For the above reaction, one can expect the change of the reaction equilibrium constant (based either on mole-fraction or molal-concentration scale) with pressure at constant temperature to be:

$$\Delta\bar{V} = \sigma\bar{V}_S + \tau\bar{V}_T - \alpha\bar{V}_A - \beta\bar{V}_B$$

The matter is complicated as partial molar volume is itself dependent on pressure.

Effect of Isotopic Substitution

Isotopic substitution can lead to changes in the values of equilibrium constants, especially if hydrogen is replaced by deuterium (or tritium). This *equilibrium isotope effect* is analogous to the kinetic isotope effect on rate constants, and is primarily due to the change in zero-point vibrational energy of H–X bonds due to the change in mass upon isotopic substitution. The zero-point energy is inversely proportional to the square root of the mass of the vibrating hydrogen atom, and will therefore be smaller for a D–X bond that for an H–X bond.

An example is a hydrogen atom abstraction reaction R' + H–R ⇌ R'–H + R with equilibrium constant K_H, where R' and R are organic radicals such that R' forms a stronger bond to hydrogen than does R. The decrease in zero-point energy due to deuterium substitution will then be more important for R'–H than for R–H, and R'–D will be stabilized more than R–D, so that the equilibrium constant K_D for R' + D–R ⇌ R'–D + R is greater than K_H.

Similar effects occur in solution for acid dissociation constants (K_a) which describe the transfer of H^+ or D^+ from a weak aqueous acid to a solvent molecule: $HA + H_2O = H_3O^+ + A^-$ or $DA + D_2O \rightleftharpoons D_3O^+ + A^-$. The deuterated acid is studied in heavy water, since if it were dissolved in ordinary water the deuterium would rapidly exchange with hydrogen in the solvent.

The product species H_3O^+ (or D_3O^+) is a stronger acid than the solute acid, so that it dissociates more easily and its H–O (or D–O) bond is weaker than the H–A (or D–A) bond of the solute acid. The decrease in zero-point energy due to isotopic substitution is therefore less important in D_3O^+ than in DA so that $K_D < K_H$, and the deuterated acid in D_2O is weaker than the non-deuterated acid in H_2O. In many cases the difference of logarithmic constants $pK_D - pK_H$ is about 0.6, so that the pD corresponding to 50% dissociation of the deuterated acid is about 0.6 units higher than the pH for 50% disssociation of the non-deuterated acid.

Stability Constants of Complexes

A stability constant (formation constant, binding constant) is an equilibrium constant for the formation of a complex in solution. It is a measure of the strength of the interaction between the reagents that come together to form the complex. There are two main kinds of complex: compounds formed by the interaction of a metal ion with a ligand and supramolecular complexes, such as host–guest complexes and complexes of anions. The stability constant(s) provide the information required to calculate the concentration(s) of the complex(es) in solution. There are many areas of application in chemistry, biology and medicine.

The formation of a complex between a metal ion, M, and a ligand, L, is in fact usually a substitution reaction. For example, in aqueous solutions, metal ions will be present as aqueous ions, so the reaction for the formation of the first complex could be written as:

$$[M(H_2O)_n] + L \rightleftharpoons [M(H_2O)_{n-1}L] + H_2O$$

The equilibrium constant for this reaction is given by,

$$\beta' = \frac{[M(H_2O)_{n-1}L][H_2O]}{[M(H_2O)_n][L]}$$

[L] should be read as "the concentration of L" and likewise for the other terms in square brackets. The expression can be greatly simplified by removing those terms which are constant. The number of water molecules attached to each metal ion is constant. In dilute solutions the concentration of water is effectively constant. The expression becomes:

$$\beta = \frac{[ML]}{[M][L]}.$$

Following this simplification a general definition can be given, for the general equilibrium,

$$p\,M + q\,L \ldots \rightleftharpoons M_pL_q\ldots$$

$$\beta_{pq\ldots} = \frac{[M_pL_q\ldots]}{[M]^p[L]^q\ldots}$$

The definition can easily be extended to include any number of reagents. The reagents need not always be a metal and a ligand but can be any species which form a complex. Stability constants defined in this way, are *association* constants. This can lead to some confusion as pK_a values are *dissociation* constants. In general purpose computer programs it is customary to define all constants as association constants. The relationship between the two types of constant is given in association and dissociation constants.

Stepwise and Cumulative Constants

A cumulative or overall constant, given the symbol β, is the constant for the formation of a complex from reagents. For example, the cumulative constant for the formation of ML_2 is given by:

$$M + 2L \rightleftharpoons ML_2;\ \beta_{12} = \frac{[ML_2]}{[M][L]^2}$$

The stepwise constants, K_1 and K_2 refer to the formation of the complexes one step at a time.

$$M + L \rightleftharpoons ML; \qquad K_1 = \frac{[ML]}{[M][L]}$$

$$ML + L \rightleftharpoons ML_2; \quad K_2 = \frac{[ML_2]}{[ML][L]}$$

It follows that,

$$\beta_{12} = K_1K_2$$

A cumulative constant can always be expressed as the product of stepwise constants. Conversely, any stepwise constant can be expressed as a quotient of two or more overall constants. There is no agreed notation for stepwise constants, though a symbol such as K^{L}_{ML} is sometimes found in the literature. It is best always to define each stability constant by reference to an equilibrium expression.

Hydrolysis Products

The formation of a hydroxo complex is a typical example of a hydrolysis reaction. A hydrolysis reaction is one in which a substrate reacts with water, splitting a water molecule into hydroxide and hydrogen ions. In this case the hydroxide ion then forms a complex with the substrate.

$$\mathrm{M + OH \rightleftharpoons M(OH)};\ K = \frac{[\mathrm{M(OH)}]}{[\mathrm{M}][\mathrm{OH}]}$$

In water the concentration of hydroxide is related to the concentration of hydrogen ions by the self-ionization constant, K_w.

$$K_w=[\mathrm{H^+}][\mathrm{OH^-}];\ [\mathrm{OH^-}] = K_w[\mathrm{H^+}]^{-1}$$

The expression for hydroxide concentration is substituted into the formation constant expression:

$$K = \frac{[\mathrm{M(OH)}]}{[\mathrm{M}]K_w[\mathrm{H}]^{-1}}$$

$$\beta^{*}_{1-1} = KK_w = \frac{[\mathrm{M(OH)}]}{[\mathrm{M}][\mathrm{H}]^{-1}}$$

The literature usually gives value of β^{*}.

Acid–base Complexes

A Lewis acid, A, and a Lewis base, B, can be considered to form a complex AB:

$$\mathrm{A + B \rightleftharpoons AB}\ K = \frac{[\mathrm{AB}]}{[\mathrm{A}][\mathrm{B}]}$$

There are three major theories relating to the strength of Lewis acids and bases and the interactions between them.

1. Hard and soft acid–base theory (HSAB): This is used mainly for qualitative purposes.
2. Drago and Wayland proposed a two-parameter equation which predicts the standard enthalpy of formation of a very large number of adducts quite accurately:

   $$-\Delta H^{\ominus}\ (\mathrm{A-B}) = E_A E_B + C_A C_B$$

3. Guttmann donor numbers: For bases the number is derived from the enthalpy of reaction of the base with antimony pentachloride in 1,2-Dichloroethane as solvent. For acids, an acceptor number is derived from the enthalpy of reaction of the acid with triphenylphosphine oxide.

Thermodynamics

The thermodynamics of metal ion complex formation provides much significant information. In particular it is useful in distinguishing between enthalpic and entropic effects. Enthalpic effects depend on bond strengths and entropic effects have to do with changes in the order/disorder of the solution as a whole. The chelate effect,, is best explained in terms of thermodynamics.

An equilibrium constant is related to the standard Gibbs free energy change for the reaction,

$$\Delta G^\ominus = -2.303\, RT \log_{10} \beta,$$

where R is the gas constant and T is the absolute temperature. At 25 °C, $\Delta G^\ominus = (-5.708 \text{ kJ mol}^{-1}) \log \beta$. Free energy is made up of an enthalpy term and an entropy term.

$$\Delta G^\ominus = \Delta H^\ominus - T\Delta S^\ominus$$

The standard enthalpy change can be determined by calorimetry or by using the Van 't Hoff equation, though the calorimetric method is preferable. When both the standard enthalpy change and stability constant have been determined, the standard entropy change is easily calculated from the equation above.

The fact that stepwise formation constants of complexes of the type ML_n decrease in magnitude as n increases may be partly explained in terms of the entropy factor. Take the case of the formation of octahedral complexes.

$$[M(H_2O)_m L_{n-1}] + L \rightleftharpoons [M(H_2O)_{m-1} L_n]$$

For the first step $m = 6$, $n = 1$ and the ligand can go into one of 6 sites. For the second step $m = 5$ and the second ligand can go into one of only 5 sites. This means that there is more randomness in the first step than the second one; $\Delta S^\ominus$ is more positive, so $\Delta G^\ominus$ is more negative and $\log K_1 > \log K_2$. The ratio of the stepwise stability constants can be calculated on this basis, but experimental ratios are not exactly the same because $\Delta H^\ominus$ is not necessarily the same for each step. The entropy factor is also important in the chelate effect.

Ionic Strength Dependence

The thermodynamic equilibrium constant, $K^\ominus$, for the equilibrium,

$$M + L \rightleftharpoons ML$$

It can be defined as,

$$K^\ominus = \frac{\{ML\}}{\{M\}\{L\}}$$

where {ML} is the activity of the chemical species ML etc. $K^\ominus$ is dimensionless since activity is dimensionless. Activities of the products are placed in the numerator, activities of the reactants are placed in the denominator.

Since activity is the product of concentration and activity coefficient (γ) the definition could also be written as,

$$K^{\ominus} = \frac{[ML]}{[M][L]} \times \frac{\gamma_{ML}}{\gamma_M \gamma_L} = \frac{[ML]}{[M][L]} \times \Gamma$$

where [ML] represents the concentration of ML and Γ is a quotient of activity coefficients. This expression can be generalized as,

$$\beta_{pq\dots}^{\ominus} = \frac{[M_pL_q\dots]}{[M]^p[L]^q\dots} \times \Gamma$$

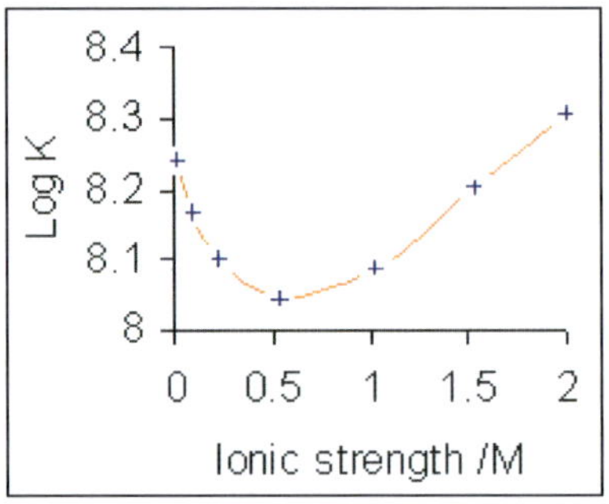

Dependence of the stability constant for formation of $[Cu(glycinate)]^+$ on ionic strength ($NaClO_4$).

To avoid the complications involved in using activities, stability constants are determined, where possible, in a medium consisting of a solution of a background electrolyte at high ionic strength, that is, under conditions in which Γ can be assumed to be always constant. For example, the medium might be a solution of 0.1 mol dm^{-3} sodium nitrate or 3 mol dm^{-3} sodium perchlorate. When Γ is constant it may be ignored and the general expression in theory is obtained.

All published stability constant values refer to the specific ionic medium used in their determination and different values are obtained with different conditions, as illustrated for the complex CuL (L = glycinate). Furthermore, stability constant values depend on the specific electrolyte used as the value of Γ is different for different electrolytes, even at the same ionic strength. There does not need to be any chemical interaction between the species in equilibrium and the background electrolyte, but such interactions might occur in particular cases. For example, phosphates form weak complexes with alkali metals, so, when determining stability constants involving phosphates, such as ATP, the background electrolyte used will be, for example, a tetralkylammonium salt. Another example involves iron(III) which forms weak complexes with halide and other anions, but not with perchlorate ions.

When published constants refer to an ionic strength other than the one required for a particular application, they may be adjusted by means of specific ion theory (SIT) and other theories.

Temperature Dependence

All equilibrium constants vary with temperature according to the Van 't Hoff equation:

$$\frac{d(\ln K)}{dT} = \frac{\Delta H_m^{\ominus}}{RT^2}$$

Alternatively,

$$\frac{d\ln K}{d(1/T)} = -\frac{\Delta H^\ominus}{R}$$

R is the gas constant and T is the thermodynamic temperature. Thus, for exothermic reactions, where the standard enthalpy change, $\Delta H^\ominus$, is negative, K decreases with temperature, but for endothermic reactions, where $\Delta H^\ominus$ is positive, K increases with temperature.

Factors affecting the Stability Constants of Complexes

Chelate Effect

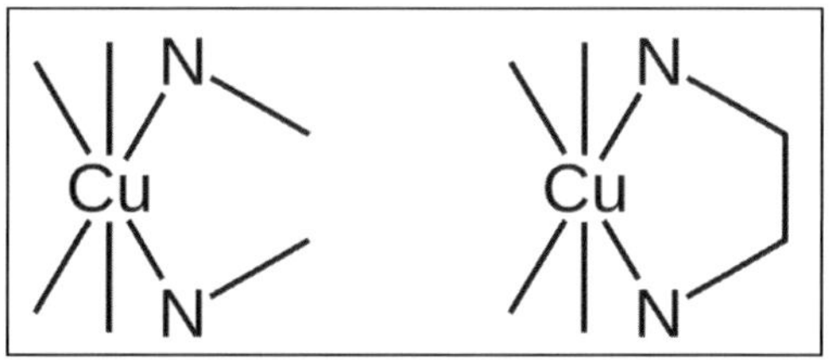

Cu^{2+} complexes with methylamine (left) end ethylene diamine (right).

Consider the two equilibria, in aqueous solution, between the copper(II) ion, Cu^{2+} and ethylenediamine (en) on the one hand and methylamine, $MeNH_2$ on the other.

$Cu^{2+} + en \rightleftharpoons [Cu(en)]^{2+}$

$Cu^{2+} + 2\ MeNH_2 \rightleftharpoons [Cu(MeNH_2)_2]^{2+}$

In $Cu^{2+} + en \rightleftharpoons [Cu(en)]^{2+}$ the bidentate ligand ethylene diamine forms a chelate complex with the copper ion. Chelation results in the formation of a five-membered ring. In $Cu^{2+} + 2\ MeNH_2 \rightleftharpoons [Cu(MeNH_2)_2]^{2+}$ the bidentate ligand is replaced by two monodentate methylamine ligands of approximately the same donor power, meaning that the enthalpy of formation of Cu–N bonds is approximately the same in the two reactions. Under conditions of equal copper concentrations and when then concentration of methylamine is twice the concentration of ethylenediamine, the concentration of the complex $Cu^{2+} + en \rightleftharpoons [Cu(en)]^{2+}$ will be greater than the concentration of the complex $Cu^{2+} + 2\ MeNH_2 \rightleftharpoons [Cu(MeNH_2)_2]^{2+}$. The effect increases with the number of chelate rings so the concentration of the EDTA complex, which has six chelate rings, is much higher than a corresponding complex with two monodentate nitrogen donor ligands and four monodentate carboxylate ligands. Thus, the phenomenon of the chelate effect is a firmly established empirical fact: under comparable conditions, the concentration of a chelate complex will be higher than the concentration of an analogous complex with monodentate ligands.

The thermodynamic approach to explaining the chelate effect considers the equilibrium constant for the reaction: the larger the equilibrium constant, the higher the concentration of the complex.

$$[Cu(en)] = \beta_{11}[Cu][en]$$

$$[Cu(MeNH_2)_2] = \beta_{12}[Cu][MeNH_2]^2$$

When the analytical concentration of methylamine is twice that of ethylenediamine and the concentration of copper is the same in both reactions, the concentration $[Cu(en)]^{2+}$ is much higher than the concentration $[Cu(MeNH_2)_2]^{2+}$ because $\beta_{11} \gg \beta_{12}$.

The difference between the two stability constants is mainly due to the difference in the standard entropy change, $\Delta S^\ominus$. In equation $Cu^{2+} + en \rightleftharpoons [Cu(en)]^{2+}$ there are two particles on the left and one on the right, whereas in equation $Cu^{2+} + 2\ MeNH_2 \rightleftharpoons [Cu(MeNH_2)_2]^{2+}$ there are three particles on the left and one on the right. This means that less entropy of disorder is lost when the chelate complex is formed than when the complex with monodentate ligands is formed. This is one of the factors contributing to the entropy difference. Other factors include solvation changes and ring formation.

Table: Some experimental data to illustrate the chelate effect.

Equilibrium	log β	$\Delta G^\ominus$/kJ mol^{-1}	$\Delta H^\ominus$/kJ mol^{-1}	$-T\Delta S^\ominus$/kJ mol^{-1}
$Cd^{2+} + 4\ MeNH_2 \rightleftharpoons Cd(MeNH_2)^{2+}_4$	6.55	−37.4	−57.3	19.9
$Cd^{2+} + 2\ en \rightleftharpoons Cd(en)^{2+}_2$	10.62	−60.67	−56.48	−4.19

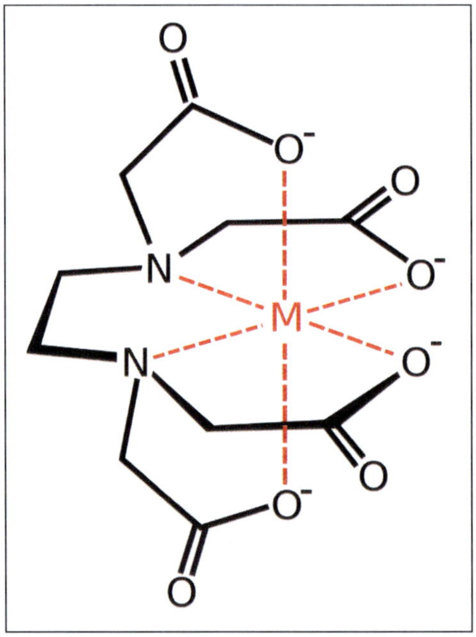

An EDTA complex.

These data show that the standard enthalpy changes are indeed approximately equal for the two reactions and that the main reason why the chelate complex is so much more stable is that the standard entropy term is much less unfavourable, indeed, it is favourable in this instance. In general it is difficult to account precisely for thermodynamic values in terms of changes in solution at the molecular level, but it is clear that the chelate effect is predominantly an effect of entropy.

The chelate effect increases as the number of chelate rings increases. For example, the complex $[Ni(dien)_2)]^{2+}$ is more stable than the complex $[Ni(en)_3)]^{2+}$; both complexes are octahedral with six nitrogen atoms around the nickel ion, but dien (diethylenetriamine, 1,4,7-triazaheptane) is a tridentate ligand and en is bidentate. The number of chelate rings is one less than the number of donor atoms in the ligand. EDTA (ethylenediaminetetracetic acid) has six donor atoms so it forms very strong complexes with five chelate rings. Ligands such as DTPA, which have eight donor atoms are used to form complexes with large metal ions such as lanthanide or actinide ions which usually form 8- or 9-coordinate complexes. 5-membered and 6-membered chelate rings give the most stable complexes. 4-membered rings are subject to internal strain because of the small inter-bond angle is the ring. The chelate effect is also reduced with 7- and 8- membered rings, because the larger rings are less rigid, so less entropy is lost in forming them.

Deprotonation of Aliphatic –OH Groups

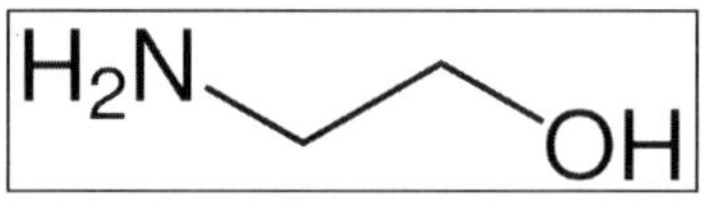

2-aminoethanol.

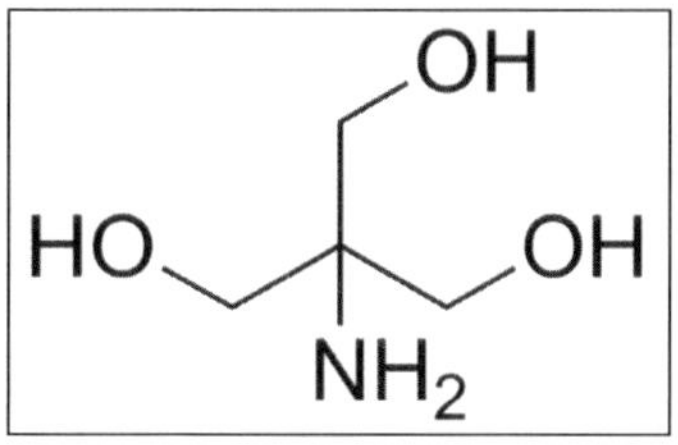

Tris.

Removal of a proton from an aliphatic –OH group is difficult to achieve in aqueous solution because the energy required for this process is rather large. Thus, ionization of aliphatic –OH groups occurs in aqueous solution only in special circumstances. One such circumstance is found with compounds containing the H_2N–C–C–OH substructure. For example, compounds containing the 2-aminoethanol substructure can form metal–chelate complexes with the deprotonated form, H_2N–C–C–O^-. The chelate effect supplies the extra energy needed to break the –OH bond.

An important example occurs with the molecule tris. This molecule should be used with caution as a buffering agent as it will form chelate complexes with ions such as Fe^{3+} and Cu^{2+}.

Macrocyclic Effect

It was found that the stability of the complex of copper(II) with the macrocyclic ligand cyclam (1,4,8,11-tetraazacyclotetradecane) was much greater than expected in comparison to the stability of the complex with the corresponding open-chain amine. This phenomenon was named "the macrocyclic effect" and it was also interpreted as an entropy effect. However, later studies suggested that both enthalpy and entropy factors were involved.

An important difference between macrocyclic ligands and open-chain (chelating) ligands is that they have selectivity for metal ions, based on the size of the cavity into which the metal ion is inserted when a complex is formed. For example, the crown ether 18-crown-6 forms much stronger complexes with the potassium ion, K^+ than with the smaller sodium ion, Na^+.

In hemoglobin an iron(II) ion is complexed by a macrocyclic porphyrin ring. It is now known that the iron(II) in hemoglobin is a low-spin complex, whereas in oxyhemoglobin it is a high-spin complex. The low-spin Fe^{2+} ion fits snugly into the cavity of the porhyrin ring, but high-spin iron(II) is significantly larger and the iron atom is forced out of the plane of the macrocyclic ligand. This effect contributes the ability of hemoglobin to bind oxygen reversibly under biological conditions.

In Vitamin B12 a cobalt(II) ion is held in a corrin ring. Chlorophyll is a macrocyclic complex of magnesium(II).

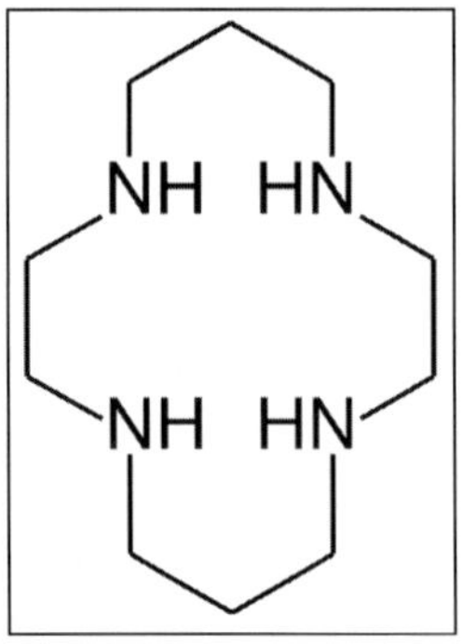

Cyclam.

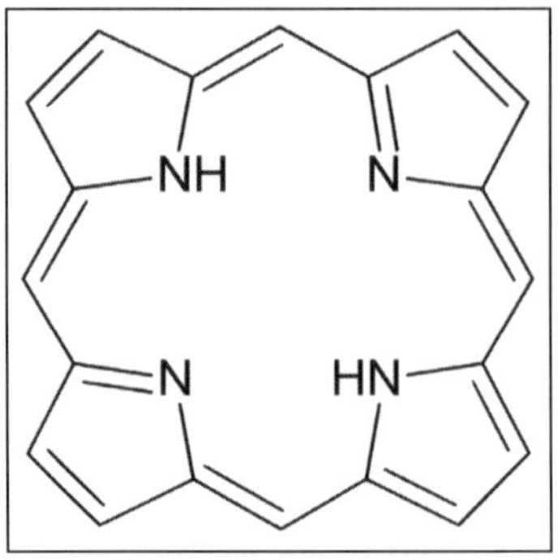

Porphine, the simplest porphyrin.

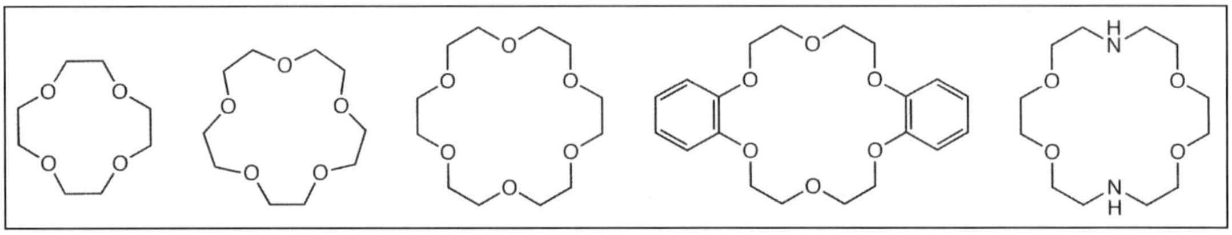

Structures of common crown ethers: 12-crown-4, 15-crown-5, 18-crown-6, dibenzo-18-crown-6, and diaza-18-crown-6.

Geometrical Factors

Successive stepwise formation constants K_n in a series such as ML_n (n = 1, 2,...) usually decrease as n increases. Exceptions to this rule occur when the geometry of the ML_n complexes is not the same for all members of the series. The classic example is the formation of the diamminesilver(I) complex $[Ag(NH_3)_2]^+$ in aqueous solution.

$$Ag^+ + NH_3 \rightleftharpoons [Ag(NH_3)]^+ \qquad K_1 = \frac{[[Ag(NH_3)]^+]}{[Ag^+][NH_3]}$$

$$Ag(NH_3)^+ + NH_3 \rightleftharpoons [Ag(NH_3)_2]^+ \qquad K_2 = \frac{[[Ag(NH_3)_2]^+]}{[[Ag(NH_3)]^+][NH_3]}$$

In this case, $K_2 > K_1$. The reason for this is that, in aqueous solution, the ion written as Ag^+ actually exists as the four-coordinate tetrahedral aqua species $[Ag(OH_2)_4]^+$. The first step is then a substitution reaction involving the displacement of a bound water molecule by ammonia forming the tetrahedral complex $[Ag(NH_3)(OH_2)_3]^+$ (commonly abbreviated as $[Ag(NH_3)]^+$). In the second step, the aqua ligands are lost to form a linear, two-coordinate product $[H_3N–Ag–NH_3]^+$. Examination of the thermodynamic data shows that both enthalpy and entropy effects determine the result.

Equilibrium	$\Delta H^\ominus$/kJ mol^{-1}	$\Delta S^\ominus$/J K^{-1} mol^{-1}
$Ag^+ + NH_3 \rightleftharpoons [Ag(NH_3)]^+$	−21.4	8.66
$[Ag(NH_3)]^+ + NH_3 \rightleftharpoons [Ag(NH_3)_2]^+$	−35.2	-61.26

Other examples exist where the change is from octahedral to tetrahedral, as in the formation of $[CoCl_4]^{2-}$ from $[Co(H_2O)_6]^{2+}$.

Classification of Metal Ions

Ahrland, Chatt and Davies proposed that metal ions could be described as class A if they formed stronger complexes with ligands whose donor atoms are nitrogen, oxygen or fluorine than with ligands whose donor atoms are phosphorus, sulfur or chlorine and class B if the reverse is true. For example, Ni^{2+} forms stronger complexes with amines than with phosphines, but Pd^{2+} forms stronger complexes with phosphines than with amines. Later, Pearson proposed the theory of hard and soft acids and bases (HSAB theory). In this classification, class A metals are hard acids and class B metals are soft acids. Some ions, such as copper(I), are classed as borderline. Hard acids form stronger complexes with hard bases than with soft bases. In general terms hard–hard interactions are predominantly electrostatic in nature whereas soft–soft interactions are predominantly covalent in nature. The HSAB theory, though useful, is only semi-quantitative.

The hardness of a metal ion increases with oxidation state. An example of this effect is given by the fact that Fe^{2+} tends to form stronger complexes with *N*-donor ligands than with *O*-donor ligands, but the opposite is true for Fe^{3+}.

Effect of Ionic Radius

The Irving–Williams series refers to high-spin, octahedral, divalent metal ion of the first transition series. It places the stabilities of complexes in the order:

$$Mn < Fe < Co < Ni < Cu > Zn$$

This order was found to hold for a wide variety of ligands. There are three strands to the explanation of the series.

1. The ionic radius is expected to decrease regularly for Mn^{2+} to Zn^{2+}. This would be the normal periodic trend and would account for the general increase in stability.

2. The crystal field stabilisation energy (CFSE) increases from zero for manganese(II) to a maximum at nickel(II). This makes the complexes increasingly stable. CFSE returns to zero for zinc(II).

3. Although the CFSE for copper(II) is less than for nickel(II), octahedral copper(II) complexes are subject to the Jahn–Teller effect which results in a complex having extra stability.

Another example of the effect of ionic radius the steady increase in stability of complexes with a given ligand along the series of trivalent lanthanide ions, an effect of the well-known lanthanide contraction.

Applications

Stability constant values are exploited in a wide variety of applications. Chelation therapy is used in the treatment of various metal-related illnesses, such as iron overload in β-thalassemia sufferers who have been given blood transfusions. The ideal ligand binds to the target metal ion and not to others, but this degree of selectivity is very hard to achieve. The synthetic drug deferiprone achieves selectivity by having two oxygen donor atoms so that it binds to Fe^{3+} in preference to any of the other divalent ions that are present in the human body, such as Mg^{2+}, Ca^{2+} and Zn^{2+}. Treatment of poisoning by ions such as Pb^{2+} and Cd^{2+} is much more difficult since these are both divalent ions and selectivity is harder to accomplish. Excess copper in Wilson's disease can be removed by penicillamine or Triethylene tetramine (TETA). DTPA has been approved by the U.S. Food and Drug Administration for treatment of plutonium poisoning.

DTPA is also used as a complexing agent for gadolinium in MRI contrast enhancement. The requirement in this case is that the complex be very strong, as Gd^{3+} is very toxic. The large stability constant of the octadentate ligand ensures that the concentration of free Gd^{3+} is almost negligible, certainly well toxicity threshold. In addition the ligand occupies only 8 of the 9 coordination sites on the gadolinium ion. The ninth site is occupied by a water molecule which exchanges rapidly with the fluid surrounding it and it is this mechanism that makes the paramagnetic complex into a contrast reagent.

EDTA forms such strong complexes with most divalent cations that it finds many uses. For example, it is often present in washing powder to act as a water softener by sequestering calcium and magnesium ions.

The selectivity of macrocyclic ligands can be used as a basis for the construction of an ion selective electrode. For example, potassium selective electrodes are available that make use of the naturally occurring macrocyclic antibiotic valinomycin.

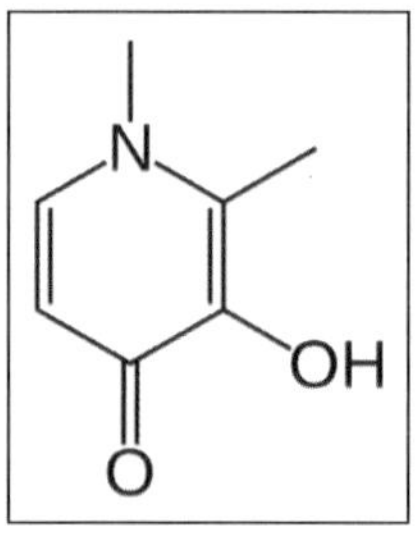

Deferiprone.

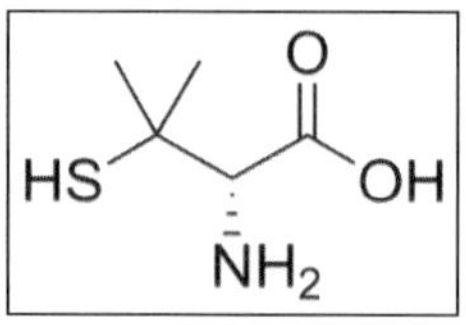

Penicillamine.

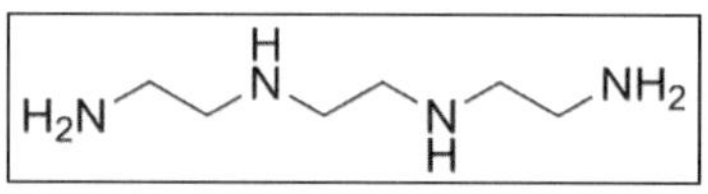

Triethylenetetramine, TETA.

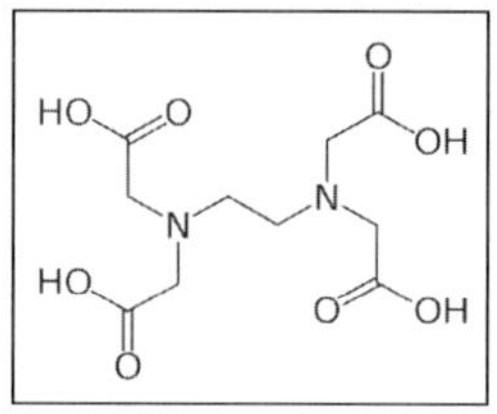

Ethylenediaminetetraacetic acid, EDTA.

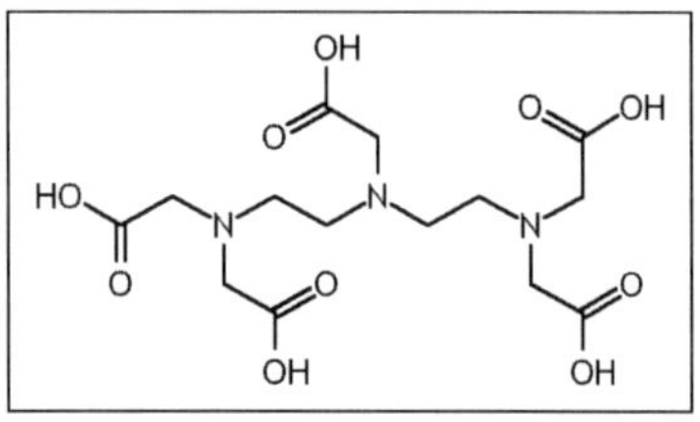

Diethylenetriaminepentaacetic acid, DTPA .

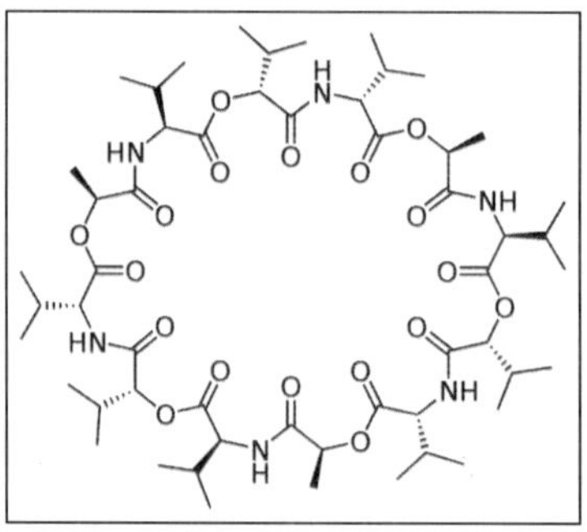

Valinomycin.

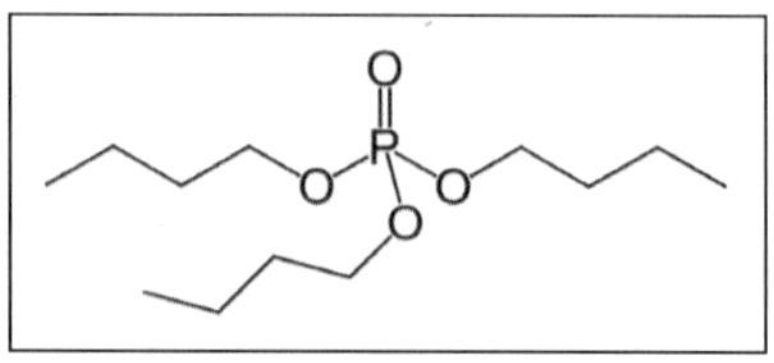

Tri-*n*-butyl phosphate.

An ion-exchange resin such as chelex 100, which contains chelating ligands bound to a polymer, can be used in water softeners and in chromatographic separation techniques. In solvent extraction the formation of electrically-neutral complexes allows cations to be extracted into organic solvents. For example, in nuclear fuel reprocessing uranium(VI) and plutonium(VI) are extracted into kerosene as the complexes $[MO_2(TBP)_2(NO_3)_2]$ (TBP = tri-*n*-butyl phosphate). In phase-transfer catalysis, a substance which is insoluble in an organic solvent can be made soluble by addition of a suitable ligand. For example, potassium permanganate oxidations can be achieved by adding a catalytic quantity of a crown ether and a small amount of organic solvent to the aqueous reaction mixture, so that the oxidation reaction occurs in the organic phase.

In all these examples, the ligand is chosen on the basis of the stability constants of the complexes formed. For example, TBP is used in nuclear fuel reprocessing because (among other reasons) it forms a complex strong enough for solvent extraction to take place, but weak enough that the complex can be destroyed by nitric acid to recover the uranyl cation as nitrato complexes, such as $[UO_2(NO_3)_4]^{2-}$ back in the aqueous phase.

Supramolecular Complexes

Supramolecular complexes are held together by hydrogen bonding, hydrophobic forces, van der Waals forces, π-π interactions, and electrostatic effects, all of which can be described as noncovalent bonding. Applications include molecular recognition, host–guest chemistry and anion sensors.

A typical application in molecular recognition involved the determination of formation constants for complexes formed between a tripodal substituted urea molecule and various saccharides. The study was carried out using a non-aqueous solvent and NMR chemical shift measurements. The object was to examine the selectivity with respect to the saccharides.

An example of the use of supramolecular complexes in the development of chemosensors is provided by the use of transition-metal ensembles to sense for ATP.

Anion complexation can be achieved by encapsulating the anion in a suitable cage. Selectivity can be engineered by designing the shape of the cage. For example, dicarboxylate anions could be encapsulated in the ellipsoidal cavity in a large macrocyclic structure containing two metal ions.

Experimental Methods

The method developed by Bjerrum is still the main method in use today, though the precision of the measurements has greatly increased. Most commonly, a solution containing the metal ion and the ligand in a medium of high ionic strength is first acidified to the point where the ligand is fully protonated. This solution is then titrated, often by means of a computer-controlled auto-titrator, with a solution of CO_2-free base. The concentration, or activity, of the hydrogen ion is monitored by means of a glass electrode. The data set used for the calculation has three components: a statement defining the nature of the chemical species that will be present, called the model of the system, details concerning the concentrations of the reagents used in the titration, and finally the experimental measurements in the form of titre and pH (or emf) pairs.

It is not always possible to use a glass electrode. If that is the case, the titration can be monitored by other types of measurement. Absorbance spectra, fluorescence spectra and NMR spectra are the most commonly used alternatives. Current practice is to take absorbance or fluorescence measurements at a range of wavelengths and to fit these data simultaneously. Various NMR chemical shifts can also be fitted together.

The chemical model will include values of the protonation constants of the ligand, which will have been determined in separate experiments, a value for log K_w and estimates of the unknown stability constants of the complexes formed. These estimates are necessary because the calculation uses a non-linear least-squares algorithm. The estimates are usually obtained by reference to a chemically similar system. The stability constant databases can be very useful in finding published stability constant values for related complexes.

In some simple cases the calculations can be done in a spreadsheet. Otherwise, the calculations are performed with the aid of a general-purpose computer programs. The most frequently used programs are:

- Potentiometric and spectrophotometric data: PSEQUAD.
- Potentiometric data: HYPERQUAD, BEST.
- Spectrophotometric data: HypSpec, SQUAD, SPECFIT, ReactLab EQUILIBRIA.
- NMR data HypNMR, WINEQNMR2

In biochemistry, formation constants of adducts may be obtained from isothermal titration calorimetry (ITC) measurements. This technique yields both the stability constant and the standard enthalpy change for the equilibrium. It is mostly limited, by availability of software, to complexes of 1:1 stoichiometry.

Acid Dissociation Constant

An acid dissociation constant, K_a (also known as acidity constant, or acid-ionization constant) is a quantitative measure of the strength of an acid in solution. It is the equilibrium constant for a chemical reaction,

$$\mathrm{HA} \rightleftharpoons \mathrm{A}^{-} + \mathrm{H}^{+}$$

It is known as dissociation in the context of acid–base reactions. The chemical species HA is an acid that dissociates into A^-, the conjugate base of the acid and a hydrogen ion, H^+. The system is said to be in equilibrium when the concentrations of its components will not change with the passing of time, because both forward and backward reactions are occurring at the same rate.

The dissociation constant is defined by,

$$pK_a = \log_{10} \frac{[\mathrm{HA}]}{[\mathrm{A}^{-}][\mathrm{H}+]}$$

where quantities in square brackets represent the concentrations of the species at equilibrium.

The acid dissociation constant for an acid is a direct consequence of the underlying thermodynamics of the dissociation reaction; the pK_a value is directly proportional to the standard Gibbs free energy change for the reaction. The value of the pK_a changes with temperature and can be understood qualitatively based on Le Chatelier's principle: when the reaction is endothermic, K_a increases and pK_a decreases with increasing temperature; the opposite is true for exothermic reactions.

The value of pK_a also depends on molecular structure of the acid in many ways. For example, Pauling proposed two rules: one for successive pK_a of polyprotic acids, and one to estimate the pK_a of oxyacids based on the number of =O and –OH groups. Other structural factors that influence the magnitude of the acid dissociation constant include inductive effects, mesomeric effects, and hydrogen bonding. Hammett type equations have frequently been applied to the estimation of pK_a.

The quantitative behavior of acids and bases in solution can be understood only if their pK_a values are known. In particular, the pH of a solution can be predicted when the analytical concentration and pK_a values of all acids and bases are known; conversely, it is possible to calculate the equilibrium concentration of the acids and bases in solution when the pH is known. These calculations find application in many different areas of chemistry, biology, medicine, and geology. For example, many compounds used for medication are weak acids or bases, and a knowledge of the pK_a values, together with the water–octanol partition coefficient, can be used for estimating the extent to which the compound enters the blood stream. Acid dissociation constants are also essential in aquatic chemistry and chemical oceanography, where the acidity of water plays a fundamental role. In living organisms, acid–base homeostasis and enzyme kinetics are dependent on the pK_a values of the many acids and bases present in the cell and in the body. In chemistry, a knowledge of pK_a values is necessary for the preparation of buffer solutions and is also a prerequisite for a quantitative understanding of the interaction between acids or bases and metal ions to form complexes. Experimentally, pK_a values can be determined by potentiometric (pH) titration, but for values of pK_a less than about 2 or more than about 11, spectrophotometric or NMR measurements may be required due to practical difficulties with pH measurements.

According to Arrhenius's original definition, an acid is a substance that dissociates in aqueous solution, releasing the hydrogen ion H^+ (a proton):

$$HA \rightleftharpoons A^- + H^+.$$

The equilibrium constant for this dissociation reaction is known as a dissociation constant. The liberated proton combines with a water molecule to give a hydronium (or oxonium) ion H_3O^+ (naked protons do not exist in solution), and so Arrhenius later proposed that the dissociation should be written as an acid–base reaction:

$$HA + H_2O \rightleftharpoons A^- + H_3O^+.$$

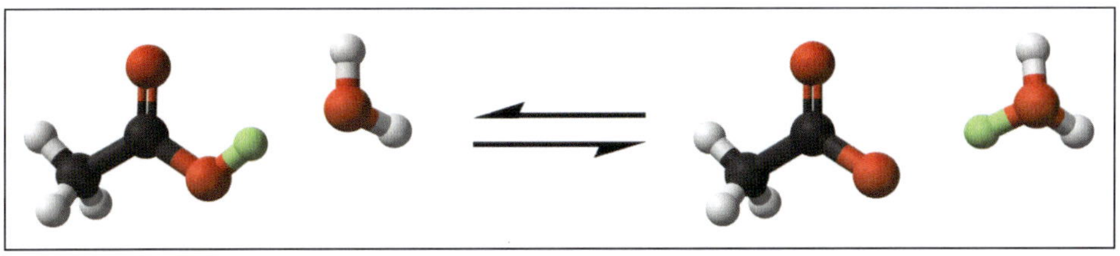

Acetic acid, a weak acid, donates a proton (hydrogen ion, highlighted in green) to water in an equilibrium reaction to give the acetate ion and the hydronium ion. Red: oxygen, black: carbon, white: hydrogen.

Bronsted and Lowry generalised this further to a proton exchange reaction:

$$\text{acid} + \text{base} \rightleftharpoons \text{conjugate base} + \text{conjugate acid}.$$

The acid loses a proton, leaving a conjugate base; the proton is transferred to the base, creating a conjugate acid. For aqueous solutions of an acid HA, the base is water; the conjugate base is A^- and the conjugate acid is the hydronium ion. The Bronsted–Lowry definition applies to other solvents, such as dimethyl sulfoxide: the solvent S acts as a base, accepting a proton and forming the conjugate acid SH^+.

$$HA + S \rightleftharpoons A^- + SH^+.$$

In solution chemistry, it is common to use H^+ as an abbreviation for the solvated hydrogen ion, regardless of the solvent. In aqueous solution H^+ denotes a solvated hydronium ion rather than a proton.

The designation of an acid or base as "conjugate" depends on the context. The conjugate acid BH^+ of a base B dissociates according to,

$$BH^+ + OH^- \rightleftharpoons B + H_2O$$

which is the reverse of the equilibrium,

$$H_2O \text{ (acid)} + B \text{ (base)} \rightleftharpoons OH^- \text{ (conjugate base)} + BH^+ \text{ (conjugate acid).}$$

The hydroxide ion OH^-, a well known base, is here acting as the conjugate base of the acid water. Acids and bases are thus regarded simply as donors and acceptors of protons respectively.

A broader definition of acid dissociation includes hydrolysis, in which protons are produced by the splitting of water molecules. For example, boric acid ($B(OH)_3$) produces H_3O^+ as if it were a proton donor, but it has been confirmed by Raman spectroscopy that this is due to the hydrolysis equilibrium:

$$B(OH)_3 + 2\,H_2O \rightleftharpoons B(OH)_4^- + H_3O^+.$$

Similarly, metal ion hydrolysis causes ions such as $[Al(H_2O)_6]^{3+}$ to behave as weak acids:

$$[Al(H_2O)_6]^{3+} + H_2O \rightleftharpoons [Al(H_2O)_5(OH)]^{2+} + H_3O^+.$$

According to Lewis's original definition, an acid is a substance that accepts an electron pair to form a coordinate covalent bond.

Equilibrium Constant

An acid dissociation constant is a particular example of an equilibrium constant. The dissociation of a monoprotic acid, HA, in dilute solution can be written as:

$$HA \rightleftharpoons A^- + H^+$$

The thermodynamic equilibrium constant, $K^\ominus$ can be defined by,

$$K^\ominus = \frac{\{A^-\}\{H+\}}{\{HA\}}$$

where {X} represents the activity, at equilibrium, of the chemical species X. $K^\ominus$ is dimensionless since activity is dimensionless. Activities of the products of dissociation are placed in the numerator, activities of the reactants are placed in the denominator.

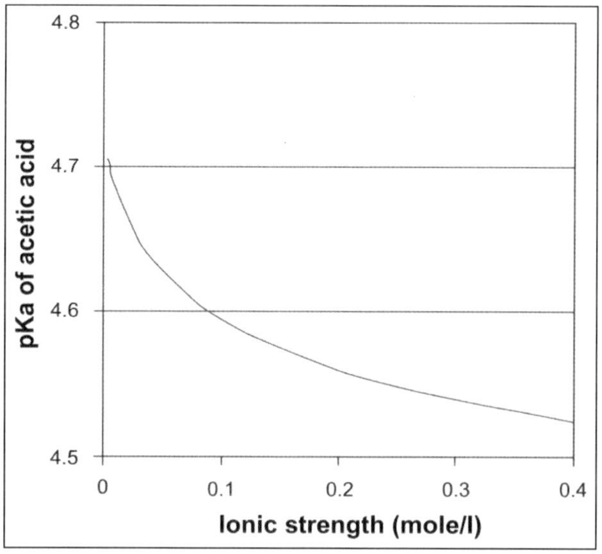

Variation of pK_a of acetic acid with ionic strength.

Since activity is the product of concentration and activity coefficient (γ) the definition could also be written as,

$$K^{\ominus} = \frac{\{A^-\}\{H+\}}{\{HA\}} \times \Gamma, \quad \Gamma = \frac{\gamma_{A^-}\,\gamma_{H^+}}{\gamma_{HA}}$$

where [HA] represents the concentration of HA and Γ is a quotient of activity coefficients.

To avoid the complications involved in using activities, dissociation constants are determined, where possible, in a medium of high ionic strength, that is, under conditions in which Γ can be assumed to be always constant. For example, the medium might be a solution of 0.1 molar (M) sodium nitrate or 3 M potassium perchlorate. With this assumption,

$$K_a = \frac{K^{\ominus}}{\Gamma} = \frac{[A^-][H^+]}{[HA]}$$

$$pK_a = \log_{10}\frac{[HA]}{[A^-][H^+]}$$

It is obtained. Note, however, that all published dissociation constant values refer to the specific ionic medium used in their determination and that different values are obtained with different conditions, as shown for acetic acid in the illustration above. When published constants refer to an ionic strength other than the one required for a particular application, they may be adjusted by means of specific ion theory (SIT) and other theories.

Cumulative and Stepwise Constants

A cumulative equilibrium constant, denoted by β, is related to the product of stepwise constants, denoted by K. For a dibasic acid the relationship between stepwise and overall constants is as follows:

$$H_2A \rightleftharpoons A^{2-} + 2H^+$$

$$\beta_2 = \frac{[H_2A]}{[A^{2-}][H+]^2}$$

$$log\beta_2 = pK_{a_1} + pK_{a_2}$$

Note that in the context of metal-ligand complex formation, the equilibrium constants for the formation of metal complexes are usually defined as *association* constants. In that case, the equilibrium constants for ligand protonation are also defined as association constants. The numbering of association constants is the reverse of the numbering of dissociation constants; in this example $log\beta_1 = pK_{a_2}$.

Association and Dissociation Constants

The dissociation of a monoprotic acid can also be described as the protonation of the conjugate base of the acid:

$$A^- + H^+ \rightleftharpoons AH$$

This leads to the definition of an *association* (protonation) constant, denoted here as K_a, as:

$$K_a = \frac{[HA]}{[A^-][H^+]}$$

The *dissociation* (deprotonation) constant definition can be written as:

$$K_d = \frac{[A^-][H^+]}{[HA]}$$

The definitions show that the values of the two constants are reciprocals of each other,

$$pK_a = \log K_d$$

with polyprotic acids: the numbering of dissociation constants is the reverse of the numbering of association constants. For example, with phosphoric acid,

$$pK_{a_1} = logK_{d_3}$$

$$pK_{a_2} = logK_{d_2}$$

$$pK_{a_3} = logK_{d_1}$$

Temperature Dependence

All equilibrium constants vary with temperature according to the Van 't Hoff equation,

$$\frac{d\ln(K)}{dT} = \frac{\Delta H^\ominus}{RT^2}$$

where R is the gas constant and T is the absolute temperature. Thus, for exothermic reactions, the standard enthalpy change, $\Delta H^\ominus$, is negative and K decreases with temperature For endothermic reactions $\Delta H^\ominus$ is positive and K increases with temperature.

The standard enthalpy change for a reaction is itself a function of temperature, according to Kirchhoff's law of thermochemistry:

$$\left(\frac{\partial \Delta H}{\partial T}\right)_p = \Delta C_p$$

where ΔC_p is the heat capacity change at constant pressure. In practice $\Delta H^{\ominus}$ may be taken to be constant over a small temperature range.

Dimensionality

In the equation,

$$K_\mathrm{a} = \frac{[\mathrm{A}^-][\mathrm{H}^+]}{[\mathrm{HA}]}$$

K_a appears to have dimensions of concentration. However, since $\Delta G = -RTlnK$, the equilibrium constant, K, *cannot* have a physical dimension. This apparent paradox can be resolved in various ways.

1. Assume that the quotient of activity coefficients has a numerical value of 1, so that K and $K^{\ominus}$ have the same numerical value.
2. Express each concentration value as the ratio c/c°, where c° is the concentration in a [hypothetical] standard state, with a numerical value of 1.
3. Express the concentrations on the mole fraction scale. Since mole fraction has no dimension, the quotient of concentrations will, be a pure number.

The procedures, (1) and (2), give identical numerical values for an equilibrium constant. Furthermore, since a concentration, c_i, is simply proportional to mole fraction, x_i and density (ρ),

$$c_i = \frac{x_i \cdot \rho}{M}$$

and since the molar mass M is a constant in dilute solutions, an equilibrium constant value determined using procedure (3) will be simply proportional to the values obtained with (1) and (2).

It is common practice in biochemistry to quote a value with a dimension as, for example, "K_a = 30 mM" in order to indicate the scale, millimolar (mM) or micromolar (μM) of the concentration values used for its calculation.

Strong Acids and Bases

An acid is classified as "strong" when the concentration of its undissociated species is too low to be measured. Any acid with a pK_a value of less than −2 is more than 99.9% dissociated at low pH. This is known as solvent leveling since all such acids are "fully dissociated", regardless of their pK_a values. It is said to be fully dissociated in aqueous solution because the amount of undissociated acid, in equilibrium with the dissociation products, is the detection limit. Likewise, any base with a pK_a value larger than about 12 is more than 99% dissociated in solution, even in strongly alkaline conditions, and is classified as a strong base.

Nitric acid, with a pK value of ca. -1.7, behaves as a strong acid in aqueous solutions with a pH greater than 1. At lower pH values it behaves as a weak acid.

pK_a values for strong acids have been estimated by theoretical means. For example, the pK_a value of aqueous HCl has been estimated as −9.3.

Monoprotic Acids

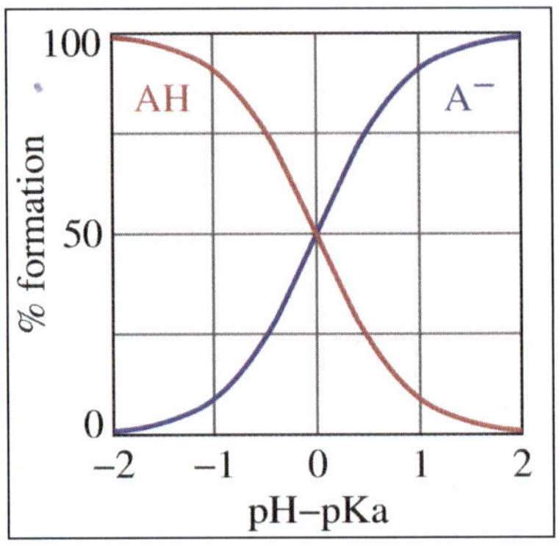

Variation of the % formation of a monoprotic acid, AH, and its conjugate base, A^-, with the difference between the pH and the pK_a of the acid.

After rearranging the expression defining K_a, and putting pH = $-\log_{10}[H^+]$, one obtains,

$$\mathrm{pH} = \mathrm{p}K_a + \log\left(\frac{[A^-]}{[HA]}\right)$$

This is the Henderson–Hasselbalch equation, from which the following conclusions can be drawn.

- At half-neutralization the ratio $[A^-]/[HA] = 1$; since log(1) = 0, the pH at half-neutralization is numerically equal to pK_a. Conversely, when pH = pK_a, the concentration of HA is equal to the concentration of A^-.
- The buffer region extends over the approximate range pK_a ± 2. Buffering is weak outside the range pK_a ± 1. At pH ≤ pK_a − 2 the substance is said to be fully protonated and at pH ≥ pK_a + 2 it is fully dissociated (deprotonated).
- If the pH is known, the ratio may be calculated. This ratio is independent of the analytical concentration of the acid.

In water, measurable pK_a values range from about −2 for a strong acid to about 12 for a very weak acid (or strong base). A buffer solution of a desired pH can be prepared as a mixture of a weak acid and its conjugate base. In practice the mixture can be created by dissolving the acid in water, and adding the requisite amount of strong acid or base. When the pK_a and analytical concentration of the acid are known, the extent of dissociation and pH of a solution of a monoprotic acid can be easily calculated using an ICE table.

Polyprotic Acids

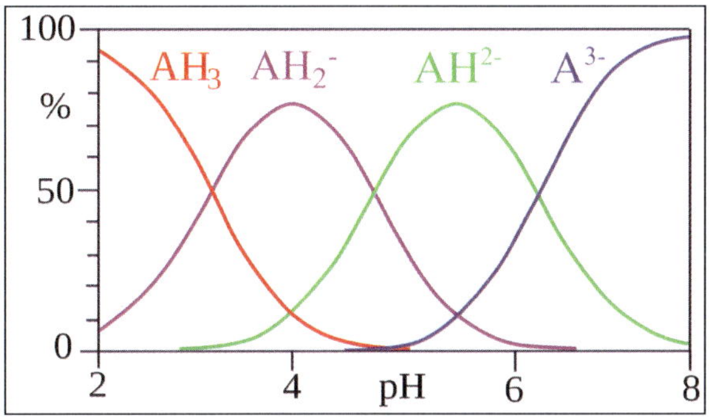

Phosphoric acid speciation.

A polyprotic acid is a compound which may lose more than 1 proton. Stepwise dissociation constants are each defined for the loss of a single proton. The constant for dissociation of the first proton may be denoted as K_{a1} and the constants for dissociation of successive protons as K_{a2}, etc. Phosphoric acid, H_3PO_4, is an example of a polyprotic acid as it can lose three protons.

Equilibrium	pK definition and value
$H_3PO_4 \rightleftharpoons H_2PO_4^- + H^+$	$pK_{a_1} = log_{10}\frac{[H_3PO_4]}{[H_2PO_4^-][H^+]} = 2.14$
$H_2PO_4^- \rightleftharpoons HPO_4^{2-} + H^+$	$pK_{a_2} = log_{10}\frac{[H_2PO_4^-]}{[HPO_4^{2-}][H^+]} = 7.2$
$HPO_4^{2-} \rightleftharpoons PO_4^{3-} + H^+$	$pK_{a_3} = log_{10}\frac{[HPO_4^{2-}]}{[PO_4^{3-}][H^+]} = 12.37$

When the difference between successive pK values is about four or more, as in this example, each species may be considered as an acid in its own right; In fact salts of $H_2PO_4^-$ may be crystallised from solution by adjustment of pH to about 5.5 and salts of HPO_4^{2-} may be crystallised from solution by adjustment of pH to about 10. The species distribution diagram shows that the concentrations of the two ions are maximum at pH 5.5 and 10.

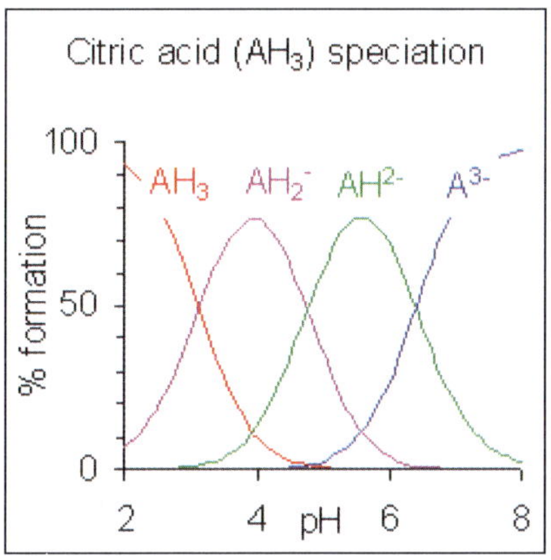

The graph shows % species formation calculated with the program HySS for a 10 millimolar solution of citric acid. pK_{a1} = 3.13, pK_{a2} = 4.76, pK_{a3} = 6.40.

When the difference between successive pK values is less than about four there is overlap between the pH range of existence of the species in equilibrium. The smaller the difference, the more the overlap. The case of citric acid is shown at the right; solutions of citric acid are buffered over the whole range of pH 2.5 to 7.5.

According to Pauling's first rule, successive pK values of a given acid increase (pK_{a2} > pK_{a1}). For oxyacids with more than one ionizable hydrogen on the same atom, the pK_a values often increase by about 5 units for each proton removed.

It can be seen in the table that the second proton is removed from a negatively charged species. Since the proton carries a positive charge extra work is needed to remove it. That why pK_{a2} is greater than pK_{a1}. pK_{a3} is greater than pK_{a2} because there is further charge separation. When an exception to Pauling's rule is found, it indicates that a major change in structure is also occurring. In the case of VO_2^+ (aq), the vanadium is octahedral, 6-coordinate, whereas vanadic acid is tetrahedral, 4-coordinate. This means that four "particles" are released with the first dissociation, but only two "particles" are released with the other dissociations, resulting in a much greater entropy contribution to the standard Gibbs free energy change for the first reaction than for the others.

Equilibrium	pK_a
$[VO_2(H_2O)_4]^+ \rightleftharpoons H_3VO_4 + H^+ + 2H_2O$	$pK_{a_1} = 4.2$
$H_3VO_4 \rightleftharpoons H_2VO^-_4 + H^+$	$pK_{a_2} = 2.60$
$H_2VO^-_4 \rightleftharpoons HVO^{2-}_4 + H^+$	$pK_{a_3} = 7.92$
$HVO^{2-}_4 \rightleftharpoons VO^{3-}_4 + H^+$	$pK_{a_4} = 13.27$

Isoelectric Point

For substances in solution the isoelectric point (pI) is defined as the pH at which the sum, weighted by charge value, of concentrations of positively charged species is equal to the weighted sum of concentrations of negatively charged species. In the case that there is one species of each type, the isoelectric point can be obtained directly from the pK values. Take the example of glycine, defined as AH. There are two dissociation equilibria to consider.

$$AH^+_2 \rightleftharpoons AH + H^+; [AH][H^+] = K_1[AH^+_2]$$

$$AH \rightleftharpoons A^- + H^+; [A^-][H^+] = K_2[AH]$$

Substitute the expression for [AH] from the second equation into the first equation:

$$[A^-][H^+]^2 = K_1K_2[AH^+_2]$$

At the isoelectric point the concentration of the positively charged species, AH_2^+, is equal to the concentration of the negatively charged species, A^-, so:

$$[H^+]^2 = K_1K_2$$

Therefore, taking cologarithms, the pH is given by,

$$pI = \frac{pK_1 + pK_2}{2}$$

where p*I* values for amino acids are listed at Proteinogenic amino acid. When more than two charged species are in equilibrium with each other a full speciation calculation may be needed.

Bases and Basicity

The equilibrium constant K_b for a base is usually defined as the *association* constant for protonation of the base, B, to form the conjugate acid, HB^+.

$$B + H_2O \rightleftharpoons HB^+ + OH^-$$

Using similar reasoning to that used before,

$$K_b = \frac{[HB^+][OH^-]}{[B]}$$

$$pK_b = -\log_{10}(K_b)$$

where K_b is related to K_a for the conjugate acid. In water, the concentration of the hydroxide ion, $[OH^-]$, is related to the concentration of the hydrogen ion by $K_w = [H^+][OH^-]$, therefore:

$$[OH^-] = \frac{K_w}{[H^+]}$$

Substitution of the expression for $[OH^-]$ into the expression for K_b gives,

$$K_b = \frac{[HB^+]K_w}{[B][H^+]} = \frac{K_w}{K_a}$$

When K_a, K_b and K_w are determined under the same conditions of temperature and ionic strength, it follows, taking cologarithms, that $pK_b = pK_w - pK_a$. In aqueous solutions at 25 °C, pK_w is 13.9965. So,

$$pK_b \approx 14 - pK_a$$

with sufficient accuracy for most practical purposes. In effect there is no need to define pK_b separately from pK_a, but it is done here as often only pK_b values can be found in the older literature.

For an hydrolyzed metal ion, K_b can also be defined as a stepwise dissociation constant:

$$M_p(OH)_q \leftrightharpoons M_p(OH)_{q-1} + OH$$

$$K_b = \frac{[M_p(OH)_{q-1}][OH]}{[M_p(OH)_q]}$$

This is the reciprocal of a association constant for formation of the complex.

Because the relationship $pK_b = pK_w - pK_a$ holds only in aqueous solutions (though analogous relationships apply for other solvents), subdisciplines of chemistry like organic chemistry that usually deal with nonaqueous solutions generally do not use pK_b as a measure of basicity. Instead, the pK_a of the conjugate acid, denoted by pK_{aH}, is quoted when basicity needs to be quantified. A higher value for pK_{aH} corresponds to a stronger base. For example, the values $pK_{aH}(C_5H_5N) = 5.25$ and $pK_{aH}((CH_3CH_2)_3N) = 10.75$ indicate that triethylamine is a stronger base than pyridine.

Amphoteric Substances

An amphoteric substance is one that can act as an acid or as a base, depending on pH. Water () is amphoteric. Another example of an amphoteric molecule is the bicarbonate ion HCO^-_3 that is the conjugate base of the carbonic acid molecule H_2CO_3 in the equilibrium,

$$H_2CO_3 + H_2O \rightleftharpoons HCO^-_3 + H_3O^+$$

but also the conjugate acid of the carbonate ion CO^{2-}_3 in (the reverse of) the equilibrium,

$$HCO^-_3 + OH^- \rightleftharpoons CO^{2-}_3 + H_2O.$$

Carbonic acid equilibria are important for acid–base homeostasis in the human body.

An amino acid is also amphoteric with the added complication that the neutral molecule is subject to an internal acid–base equilibrium in which the basic amino group attracts and binds the proton from the acidic carboxyl group, forming a zwitterion.

$$NH_2CHRCO_2H \rightleftharpoons NH^+_3CHRCO^-_2$$

At pH less than about 5 both the carboxylate group and the amino group are protonated. As pH increases the acid dissociates according to:

$$NH^+_3CHRCO_2H \rightleftharpoons NH^+_3CHRCO^-_2 + H^+$$

At high pH a second dissociation may take place.

$$NH^+_3CHRCO^-_2 \rightleftharpoons NH_2CHRCO^-_2 + H^+$$

Thus the amino acid molecule is amphoteric because it may either be protonated or deprotonated.

Water Self-ionization

The water molecule may either gain or lose a proton. It is said to be amphiprotic. The ionization equilibrium can be written,

$$H_2O \rightleftharpoons OH^- + H^+$$

where in aqueous solution H^+ denotes a solvated proton. Often this is written as the hydronium ion H_3O^+, but this formula is not exact because in fact there is solvation by more than one water molecule and species such as $H_5O_2^+$, $H_7O_3^+$ and $H_9O_4^+$ are also present.

The equilibrium constant is given by,

$$K_a = \frac{[H^+][OH^-]}{[H_2O]}$$

with solutions in which the solute concentrations are not very high, the concentration $[H_2O]$ can be assumed to be constant, regardless of solute(s); this expression may then be replaced by:

$$K_w = [H^+][OH^-]$$

The self-ionization constant of water, K_w, is thus just a special case of an acid dissociation constant. A logarithmic form analogous to pK_a may also be defined:

$$pK_w = -\log_{10}(K_w)$$

pK_w values for pure water at various temperatures											
T **(°C)**	0	5	10	15	20	25	30	35	40	45	50
pK_w	14.943	14.734	14.535	14.346	14.167	13.997	13.830	13.680	13.535	13.396	13.262

These data can be fitted to a parabola with:

$$pK_w = 14.94 - 0.04209T + 0.0001718T^2$$

From this equation, $pK_w = 14$ at 24.87 °C. At that temperature both hydrogen and hydroxide ions have a concentration of 10^{-7} M.

Acidity in Nonaqueous Solutions

A solvent will be more likely to promote ionization of a dissolved acidic molecule in the following circumstances:

1. It is a protic solvent, capable of forming hydrogen bonds.
2. It has a high donor number, making it a strong Lewis base.
3. It has a high dielectric constant (relative permittivity), making it a good solvent for ionic species.

pK_a values of organic compounds are often obtained using the aprotic solvents dimethyl sulfoxide (DMSO) and acetonitrile (ACN).

Solvent properties at 25 °C		
Solvent	Donor number	Dielectric constant
Acetonitrile	14	37
Dimethylsulfoxide	30	47
Water	18	78

DMSO is widely used as an alternative to water because it has a lower dielectric constant than water, and is less polar and so dissolves non-polar, hydrophobic substances more easily. It has a measurable pK_a range of about 1 to 30. Acetonitrile is less basic than DMSO, and, so, in general, acids are weaker and bases are stronger in this solvent. Some pK_a values at 25 °C for acetonitrile (ACN) and dimethyl sulfoxide (DMSO) are shown in the following tables. Values for water are included for comparison.

pK_a values of acids			
$HA \rightleftharpoons A^- + H^+$	ACN	DMSO	Water
p-Toluenesulfonic acid	8.5	0.9	*Strong*
2,4-Dinitrophenol	16.66	5.1	3.9
Benzoic acid	21.51	11.1	4.2
Acetic acid	23.51	12.6	4.756
Phenol	29.14	18.0	9.99
$BH^+ \rightleftharpoons B + H^+$	ACN	DMSO	Water
Pyrrolidine	19.56	10.8	11.4
Triethylamine	18.82	9.0	10.72
Proton sponge	18.62	7.5	12.1
Pyridine	12.53	3.4	5.2
Aniline	10.62	3.6	4.6

Ionization of acids is less in an acidic solvent than in water. For example, hydrogen chloride is a weak acid when dissolved in acetic acid. This is because acetic acid is a much weaker base than water.

$$HCl + CH_3CO_2H \rightleftharpoons Cl^- + CH_3C(OH)^+_2$$

acid + base ⇌ conjugate base + conjugate acid

Compare this reaction with what happens when acetic acid is dissolved in the more acidic solvent pure sulfuric acid:

$$H_2SO_4 + CH_3CO_2H \rightleftharpoons HSO^-_4 + CH_3C(OH)^+_2$$

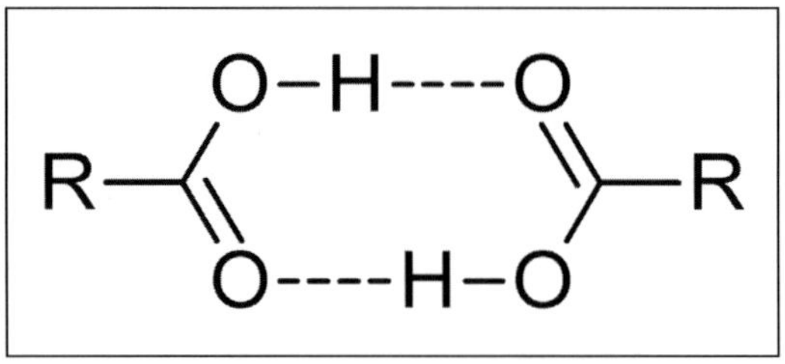

Dimerization of a carboxylic acid.

The unlikely geminal diol species $CH_3C(OH)^+_2$ is stable in these environments. For aqueous solutions the pH scale is the most convenient acidity function. Other acidity functions have been proposed for non-aqueous media, the most notable being the Hammett acidity function, H_o, for superacid media and its modified version H^- for superbasic media.

In aprotic solvents, oligomers, such as the well-known acetic acid dimer, may be formed by hydrogen bonding. An acid may also form hydrogen bonds to its conjugate base. This process, known as homoconjugation, has the effect of enhancing the acidity of acids, lowering their effective pK_a values, by stabilizing the conjugate base. Homoconjugation enhances the proton-donating power of toluenesulfonic acid in acetonitrile solution by a factor of nearly 800.

In aqueous solutions, homoconjugation does not occur, because water forms stronger hydrogen bonds to the conjugate base than does the acid.

Mixed Solvents

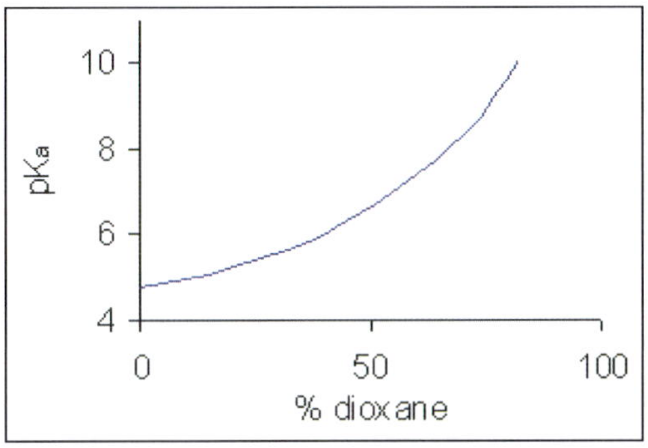

pK_a of acetic acid in dioxane/water mixtures.

When a compound has limited solubility in water it is common practice (in the pharmaceutical industry, for example) to determine pK_a values in a solvent mixture such as water/dioxane or water/methanol, in which the compound is more soluble. The pK_a value rises steeply with increasing percentage of dioxane as the dielectric constant of the mixture is decreasing.

A pK_a value obtained in a mixed solvent cannot be used directly for aqueous solutions. The reason for this is that when the solvent is in its standard state its activity is *defined* as one. For example, the standard state of water:dioxane mixture with 9:1 mixing ratio is precisely that solvent mixture, with no added solutes. To obtain the pK_a value for use with aqueous solutions it has to be extrapolated to zero co-solvent concentration from values obtained from various co-solvent mixtures.

These facts are obscured by the omission of the solvent from the expression that is normally used to define pK_a, but pK_a values obtained in a given mixed solvent can be compared to each other, giving relative acid strengths. The same is true of pK_a values obtained in a particular non-aqueous solvent such a DMSO.

A universal, solvent-independent, scale for acid dissociation constants has not been developed, since there is no known way to compare the standard states of two different solvents.

Factors that Affect pK_a Values

Pauling's second rule is that the value of the first pK_a for acids of the formula $XO_m(OH)_n$ depends primarily on the number of oxo groups m, and is approximately independent of the number of hydroxy groups n, and also of the central atom X. Approximate values of pK_a are 8 for m = 0, 2 for m = 1, −3 for m = 2 and < −10 for m = 3. Alternatively, various numerical formulas have been proposed including pK_a = 8 − 5m (known as Bell's rule), pK_a = 7 − 5m, or pK_a = 9 − 7m. The dependence on m correlates with the oxidation state of the central atom, X: the higher the oxidation state the stronger the oxyacid. For example, pK_a for HClO is 7.2, for $HClO_2$ is 2.0, for $HClO_3$ is −1 and $HClO_4$ is a strong acid (p$K_a \ll 0$). The increased acidity on adding an oxo group is due to stabilization of the conjugate base by delocalization of its negative charge over an additional oxygen atom. This rule can help assign molecular structure: for example phosphorous acid (H_3PO_3) has a pK_a near 2 suggested that the structure is $HPO(OH)_2$, as later confirmed by NMR spectroscopy, and not $P(OH)_3$ which would be expected to have a pK_a near 8.

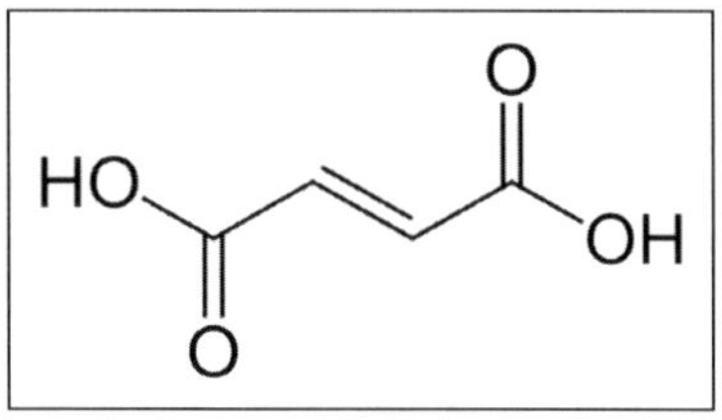

Fumaric acid.

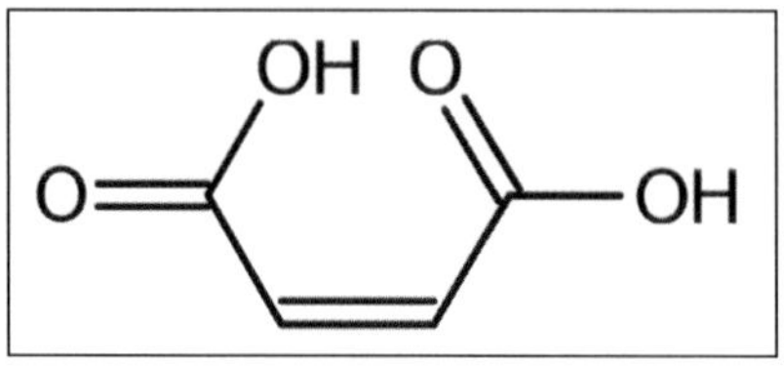

Maleic acid.

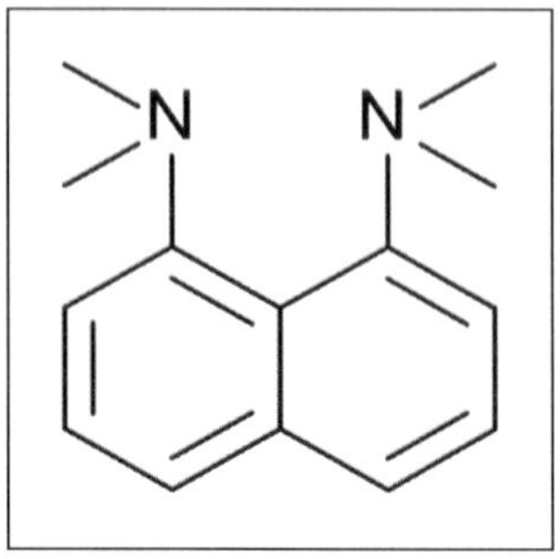

Proton sponge.

With organic acids inductive effects and mesomeric effects affect the pK_a values. A simple example is provided by the effect of replacing the hydrogen atoms in acetic acid by the more electronegative chlorine atom. The electron-withdrawing effect of the substituent makes ionisation easier, so successive pK_a values decrease in the series 4.7, 2.8, 1.4, and 0.7 when 0, 1, 2, or 3 chlorine atoms are present. The Hammett equation, provides a general expression for the effect of substituents.

$$\log(K_a) = \log(K^o_a) + \rho\sigma.$$

K_a is the dissociation constant of a substituted compound, K^o_a is the dissociation constant when the substituent is hydrogen, ρ is a property of the unsubstituted compound and σ has a particular value for each substituent. A plot of $\log(K_a)$ against σ is a straight line with intercept $\log(K^o_a)$ and slope ρ. This is an example of a linear free energy relationship as $\log(K_a)$ is proportional to the standard free energy change. Hammett originally formulated the relationship with data from benzoic acid with different substiuents in the *ortho*- and *para*- positions: some numerical values are in Hammett equation. This and other studies allowed substituents to be ordered according to their electron-withdrawing or electron-releasing power, and to distinguish between inductive and mesomeric effects.

Alcohols do not normally behave as acids in water, but the presence of a double bond adjacent to the OH group can substantially decrease the pK_a by the mechanism of keto–enol tautomerism. Ascorbic acid is an example of this effect. The diketone 2,4-pentanedione (acetylacetone) is also a weak acid because of the keto–enol equilibrium. In aromatic compounds, such as phenol, which have an OH substituent, conjugation with the aromatic ring as a whole greatly increases the stability of the deprotonated form.

Structural effects can also be important. The difference between fumaric acid and maleic acid is a classic example. Fumaric acid is (E)-1,4-but-2-enedioic acid, a *trans* isomer, whereas maleic acid is the corresponding *cis* isomer, i.e. (Z)-1,4-but-2-enedioic acid. Fumaric acid has pK_a values of approximately 3.0 and 4.5. By contrast, maleic acid has pK_a values of approximately 1.5 and 6.5. The reason for this large difference is that when one proton is removed from the *cis* isomer (maleic acid) a strong intramolecular hydrogen bond is formed with the nearby remaining carboxyl group.

This favors the formation of the maleate H^+, and it opposes the removal of the second proton from that species. In the *trans* isomer, the two carboxyl groups are always far apart, so hydrogen bonding is not observed.

Proton sponge, 1,8-bis(dimethylamino)naphthalene, has a pK_a value of 12.1. It is one of the strongest amine bases known. The high basicity is attributed to the relief of strain upon protonation and strong internal hydrogen bonding.

It turns out, these influences are more subtle than that of a dielectric medium. For example, the expected (by electronic effects of methyl substituents) and observed in gas phase order of basicity of methylamines, $Me_3N > Me_2NH > MeNH_2 > NH_3$, is changed by water to $Me_2NH > MeNH_2 > Me_3N > NH_3$. Neutral methylamine molecules are hydrogen-bonded to water molecules mainly through one acceptor, N–HOH, interaction and only occasionally just one more donor bond, $NH–OH_2$. Hence, methylamines are stabilized to about the same extent by hydration, regardless of the number of methyl groups. In stark contrast, corresponding methylammonium cations always utilize all the available protons for donor $NH–OH_2$ bonding. Relative stabilization of methylammonium ions thus decreases with the number of methyl groups explaining the order of water basicity of methylamines.

Thermodynamics

An equilibrium constant is related to the standard Gibbs energy change for the reaction, so for an acid dissociation constant,

$$\Delta G^{\ominus} = -RT \ln K_a \approx 2.303RT \, pK_a$$

where R is the gas constant and T is the absolute temperature. Note that $pK_a = -\log(K_a)$ and 2.303 ≈ ln(10). At 25 °C, $\Delta G^{\ominus}$ in $kJ \cdot mol^{-1} \approx 5.708 \, pK_a$ ($1 \, kJ \cdot mol^{-1}$ = 1000 joules per mole). Free energy is made up of an enthalpy term and an entropy term.

$$\Delta G^{\ominus} = \Delta H^{\ominus} - T\Delta S^{\ominus}$$

The standard enthalpy change can be determined by calorimetry or by using the Van 't Hoff equation, though the calorimetric method is preferable. When both the standard enthalpy change and acid dissociation constant have been determined, the standard entropy change is easily calculated from the equation above. In the following table, the entropy terms are calculated from the experimental values of pK_a and $\Delta H^{\ominus}$. The data were critically selected and refer to 25 °C and zero ionic strength, in water.

Acids					
Compound	Equilibrium	pK_a	$\Delta G^{\ominus}$ ($kJ \cdot mol^{-1}$)	$\Delta H^{\ominus}$ ($kJ \cdot mol^{-1}$)	$-T\Delta S^{\ominus}$ ($kJ \cdot mol^{-1}$)
HA = Acetic acid	$HA \rightleftharpoons H^+ + A^-$	4.756	27.147	−0.41	27.56
H_2A^+ = GlycineH$^+$	$H_2A^+ \rightleftharpoons HA + H^+$	2.351	13.420	4.00	9.419
	$HA \rightleftharpoons H^+ + A^-$	9.78	55.825	44.20	11.6
H_2A = Maleic acid	$H_2A \rightleftharpoons HA^- + H^+$	1.92	10.76	1.10	9.85
	$HA^- \rightleftharpoons H^+ + A^{2-}$	6.27	35.79	−3.60	39.4

H_3A = Citric acid	$H_3A \rightleftharpoons H_2A^- + H^+$	3.128	17.855	4.07	13.78
	$H_2A^- \rightleftharpoons HA^{2-} + H^+$	4.76	27.176	2.23	24.9
	$HA^{2-} \rightleftharpoons A^{3-} + H^+$	6.40	36.509	−3.38	39.9
H_3A = Boric acid	$H_3A \rightleftharpoons H_2A^- + H^+$	9.237	52.725	13.80	38.92
H_3A = Phosphoric acid	$H_3A \rightleftharpoons H_2A^- + H^+$	2.148	12.261	−8.00	20.26
	$H_2A^- \rightleftharpoons HA^{2-} + H^+$	7.20	41.087	3.60	37.5
	$HA^{2-} \rightleftharpoons A^{3-} + H^+$	12.35	80.49	16.00	54.49
HA^- = Hydrogen sulfate	$HA^- \rightleftharpoons A^{2-} + H^+$	1.99	11.36	−22.40	33.74
H_2A = Oxalic acid	$H_2A \rightleftharpoons HA^- + H^+$	1.27	7.27	−3.90	11.15
	$HA^- \rightleftharpoons A^{2-} + H^+$	4.266	24.351	−7.00	31.35

1. $\Delta G^\ominus \approx 2.303RT\mathrm{p}K_a$.
2. The value of ΔH and ΔG supplied using: $-T\Delta S^\ominus = \Delta G^\ominus - \Delta H^\ominus$.

Conjugate acids of bases				
Compound	Equilibrium	pK_a	$\Delta H^\ominus$ (kJ·mol^{-1})	$-T\Delta S^\ominus$ (kJ·mol^{-1})
B = Ammonia	$HB^+ \rightleftharpoons B + H^+$	9.245	51.95	0.8205
B = Methylamine	$HB^+ \rightleftharpoons B + H^+$	10.645	55.34	5.422
B = Triethylamine	$HB^+ \rightleftharpoons B + H^+$	10.72	43.13	18.06

The first point to note is that, when pK_a is positive, the standard free energy change for the dissociation reaction is also positive. Second, some reactions are exothermic and some are endothermic, but, when $\Delta H^\ominus$ is negative $T\Delta S^\ominus$ is the dominant factor, which determines that $\Delta G^\ominus$ is positive. Last, the entropy contribution is always unfavourable ($\Delta S^\ominus < 0$) in these reactions. Ions in aqueous solution tend to orient the surrounding water molecules, which orders the solution and decreases the entropy. The contribution of an ion to the entropy is the partial molar entropy which is often negative, especially for small or highly charged ions. The ionization of a neutral acid involves formation of two ions so that the entropy decreases ($\Delta S^\ominus < 0$). On the second ionization of the same acid, there are now three ions and the anion has a charge, so the entropy again decreases.

Note that the *standard* free energy change for the reaction is for the changes *from* the reactants in their standard states *to* the products in their standard states. The free energy change *at* equilibrium is zero since the chemical potentials of reactants and products are equal at equilibrium.

Experimental Determination

The experimental determination of pK_a values is commonly performed by means of titrations, in a medium of high ionic strength and at constant temperature. A typical procedure would be as follows. A solution of the compound in the medium is acidified with a strong acid to the point where the compound is fully protonated. The solution is then titrated with a strong base until all the protons have been removed. At each point in the titration pH is measured using a glass electrode and a pH meter. The equilibrium constants are found by fitting calculated pH values to the observed values, using the method of least squares.

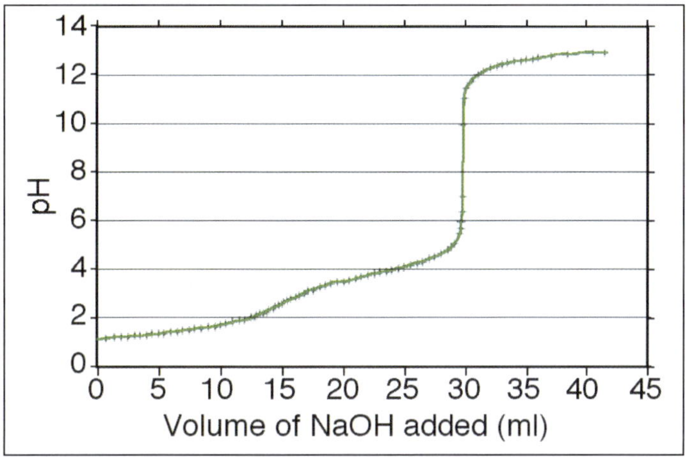

A calculated titration curve of oxalic acid titrated with a solution of sodium hydroxide.

The total volume of added strong base should be small compared to the initial volume of titrand solution in order to keep the ionic strength nearly constant. This will ensure that pK_a remains invariant during the titration.

Oxalic acid has pK_a values of 1.27 and 4.27. Therefore, the buffer regions will be centered at about pH 1.3 and pH 4.3. The buffer regions carry the information necessary to get the pK_a values as the concentrations of acid and conjugate base change along a buffer region.

Between the two buffer regions there is an end-point, or equivalence point, at about pH 3. This end-point is not sharp and is typical of a diprotic acid whose buffer regions overlap by a small amount: $pK_{a2} - pK_{a1}$ is about three in this example. If the difference in p*K* values were about two or less, the end-point would not be noticeable. The second end-point begins at about pH 6.3 and is sharp. This indicates that all the protons have been removed. When this is so, the solution is not buffered and the pH rises steeply on addition of a small amount of strong base. However, the pH does not continue to rise indefinitely. A new buffer region begins at about pH 11 ($pK_w - 3$), which is where self-ionization of water becomes important.

It is very difficult to measure pH values of less than two in aqueous solution with a glass electrode, because the Nernst equation breaks down at such low pH values. To determine p*K* values of less than about 2 or more than about 11 spectrophotometric or NMR measurements may be used instead of, or combined with, pH measurements.

When the glass electrode cannot be employed, as with non-aqueous solutions, spectrophotometric methods are frequently used. These may involve absorbance or fluorescence measurements. In both cases the measured quantity is assumed to be proportional to the sum of contributions from each photo-active species; with absorbance measurements the Beer-Lambert law is assumed to apply.

Isothermal titration calorimetry (ITC) may be used to determine both a pK value and the corresponding standard enthalpy for acid dissociation. Software to perform the calculations is supplied by the instrument manufacturers for simple systems.

Aqueous solutions with normal water cannot be used for ^{1}H NMR measurements but heavy water, D_2O, must be used instead. ^{13}C NMR data, however, can be used with normal water and ^{1}H NMR spectra can be used with non-aqueous media. The quantities measured with NMR are time-averaged chemical shifts, as proton exchange is fast on the NMR time-scale. Other chemical shifts, such as those of ^{31}P can be measured.

Micro-constants

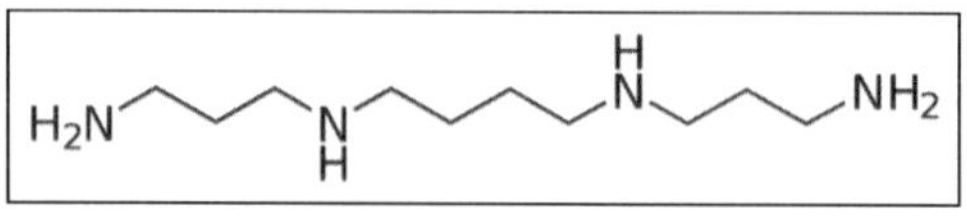

Spermine.

A base such as spermine has more than one site where protonation can occur. For example, mono-protonation can occur at a terminal $-NH_2$ group, or an internal $-NH-$ groups. The pK_a values for dissociation of spermine protonated at one or other of the sites are examples of micro-constants. They cannot be determined directly by means of pH, absorbance, fluorescence or NMR measurements. Nevertheless, the site of protonation is very important for biological function, so mathematical methods have been developed for the determination of micro-constants.

Applications and Significance

A knowledge of pK_a values is important for the quantitative treatment of systems involving acid–base equilibria in solution. Many applications exist in biochemistry; for example, the pK_a values of proteins and amino acid side chains are of major importance for the activity of enzymes and the stability of proteins. Protein pK_a values cannot always be measured directly, but may be calculated using theoretical methods. Buffer solutions are used extensively to provide solutions at or near the physiological pH for the study of biochemical reactions; the design of these solutions depends on a knowledge of the pK_a values of their components. Important buffer solutions include MOPS, which provides a solution with pH 7.2, and tricine, which is used in gel electrophoresis. Buffering is an essential part of acid base physiology including acid–base homeostasis, and is key to understanding disorders such as acid–base imbalance. The isoelectric point of a given molecule is a function of its pK values, so different molecules have different isoelectric points. This permits a technique called isoelectric focusing, which is used for separation of proteins by 2-D gel polyacrylamide gel electrophoresis.

Buffer solutions also play a key role in analytical chemistry. They are used whenever there is a need to fix the pH of a solution at a particular value. Compared with an aqueous solution, the pH of a buffer solution is relatively insensitive to the addition of a small amount of strong acid or strong base. The buffer capacity of a simple buffer solution is largest when pH = pK_a. In acid–base extraction, the efficiency of extraction of a compound into an organic phase, such as an ether, can be optimised by adjusting the pH of the aqueous phase using an appropriate buffer. At the optimum pH, the concentration of the electrically neutral species is maximised; such a species is more soluble in organic solvents having a low dielectric constant than it is in water. This technique is used for the purification of weak acids and bases.

A pH indicator is a weak acid or weak base that changes color in the transition pH range, which is approximately pK_a ± 1. The design of a universal indicator requires a mixture of indicators whose adjacent pK_a values differ by about two, so that their transition pH ranges just overlap.

In pharmacology, ionization of a compound alters its physical behavior and macro properties such as solubility and lipophilicity, log p). For example, ionization of any compound will increase the solubility in water, but decrease the lipophilicity. This is exploited in drug development to increase the concentration of a compound in the blood by adjusting the pK_a of an ionizable group.

Knowledge of pK_a values is important for the understanding of coordination complexes, which are formed by the interaction of a metal ion, M^{m+}, acting as a Lewis acid, with a ligand, L, acting as a Lewis base. However, the ligand may also undergo protonation reactions, so the formation of a complex in aqueous solution could be represented symbolically by the reaction:

$$[M(H_2O)_n]^{m+} + LH \rightleftharpoons [M(H_2O)_{n-1}L]^{(m-1)+} + H_3O^+$$

To determine the equilibrium constant for this reaction, in which the ligand loses a proton, the pK_a of the protonated ligand must be known. In practice, the ligand may be polyprotic; for example $EDTA^{4-}$ can accept four protons; in that case, all pK_a values must be known. In addition, the metal ion is subject to hydrolysis, that is, it behaves as a weak acid, so the pK values for the hydrolysis reactions must also be known.

Assessing the hazard associated with an acid or base may require a knowledge of pK_a values. For example, hydrogen cyanide is a very toxic gas, because the cyanide ion inhibits the iron-containing enzyme cytochrome c oxidase. Hydrogen cyanide is a weak acid in aqueous solution with a pK_a of about 9. In strongly alkaline solutions, above pH 11, say, it follows that sodium cyanide is "fully dissociated" so the hazard due to the hydrogen cyanide gas is much reduced. An acidic solution, on the other hand, is very hazardous because all the cyanide is in its acid form. Ingestion of cyanide by mouth is potentially fatal, independently of pH, because of the reaction with cytochrome c oxidase.

In environmental science acid–base equilibria are important for lakes and rivers; for example, humic acids are important components of natural waters. Another example occurs in chemical oceanography: in order to quantify the solubility of iron(III) in seawater at various salinities, the pK_a values for the formation of the iron(III) hydrolysis products $Fe(OH)^{2+}$, $Fe(OH)^+_2$ and $Fe(OH)_3$ were determined, along with the solubility product of iron hydroxide.

Values for Common Substances

There are multiple techniques to determine the pK_a of a chemical, leading to some discrepancies between different sources. Well measured values are typically within 0.1 units of each other. Data presented here were taken at 25 °C in water.

Chemical	Equilibrium	pK_a
BH = Adenine	$BH \rightleftharpoons B^- + H^+$	4.17
	$BH^+_2 \rightleftharpoons BH + H^+$	9.65
H_3A = Arsenic acid	$H_3A \rightleftharpoons H_2A^- + H^+$	2.22
	$H_2A^- \rightleftharpoons HA^{2-} + H^+$	6.98
	$HA^{2-} \rightleftharpoons A^{3-} + H^+$	11.53
HA = Benzoic acid	$HA \rightleftharpoons H^+ + A^-$	4.204
HA = Butyric acid	$HA \rightleftharpoons H^+ + A^-$	4.82
H_2A = Chromic acid	$H_2A \rightleftharpoons HA^- + H^+$	0.98
	$HA^- \rightleftharpoons A^{2-} + H^+$	6.5

B = Codeine	$BH^+ \rightleftharpoons B + H^+$	8.17
HA = Cresol	$HA \rightleftharpoons H^+ + A^-$	10.29
HA = Formic acid	$HA \rightleftharpoons H^+ + A^-$	3.751
HA = Hydrofluoric acid	$HA \rightleftharpoons H^+ + A^-$	3.17
HA = Hydrocyanic acid	$HA \rightleftharpoons H^+ + A^-$	9.21
HA = Hydrogen selenide	$HA \rightleftharpoons H^+ + A^-$	3.89
HA = Hydrogen peroxide (90%)	$HA \rightleftharpoons H^+ + A^-$	11.7
HA = Lactic acid	$HA \rightleftharpoons H^+ + A^-$	3.86
HA = Propionic acid	$HA \rightleftharpoons H^+ + A^-$	4.87
HA = Phenol	$HA \rightleftharpoons H^+ + A^-$	9.99
H_2A = L-(+)-Ascorbic Acid	$H_2A \rightleftharpoons HA^- + H^+$	4.17
	$HA^- \rightleftharpoons A^{2-} + H^+$	11.57

Van 't Hoff Equation

The Van 't Hoff equation relates the change in the equilibrium constant, K_{eq}, of a chemical reaction to the change in temperature, T, given the standard enthalpy change, $\Delta H^\ominus$, for the process. It was proposed by Dutch chemist Jacobus Henricus Van 't Hoff in 1884. This equation is sometimes also referred to as the Vukančić–Vuković equation.

The Van 't Hoff equation has been widely utilized to explore the changes in state functions in a thermodynamic system. The Van 't Hoff plot, which is derived from this equation, is especially effective in estimating the change in enthalpy, or total energy, and entropy, or amount of disorder, of a chemical reaction.

Equation:

Under standard conditions, the Van 't Hoff equation is,

$$\frac{d}{dT}\ln K_{\text{eq}} = \frac{\Delta H^\ominus}{RT^2}$$

where *ln* denotes natural logarithm and R is the ideal gas constant. This equation is exact at any one temperature. In practice, the equation is often integrated between two temperatures under the assumption that the reaction enthalpy ΔH is constant. Since in reality ΔH and the reaction entropy ΔS do vary with temperature for most processes, the integrated equation is only approximate.

A major use of the integrated equation is to estimate a new equilibrium constant at a new absolute temperature assuming a constant standard enthalpy change over the temperature range.

To obtain the integrated equation, it is convenient to first rewrite the Van 't Hoff equation as:

$$\frac{d\ln K_{\text{eq}}}{d\frac{1}{T}} = -\frac{\Delta H^\ominus}{R}$$

The definite integral between temperatures T_1 and T_2 is then,

$$\ln\frac{K_2}{K_1} = \frac{-\Delta H^\ominus}{R}\left(\frac{1}{T_2} - \frac{1}{T_1}\right)$$

In this equation K_1 is the equilibrium constant at absolute temperature T_1, and K_2 is the equilibrium constant at absolute temperature T_2.

Development from Thermodynamics

Combining the definition of Gibbs free energy,

$$\Delta G^\ominus = \Delta H^\ominus - T\Delta S^\ominus,$$

where S is the entropy of the system, and the Gibbs free energy isotherm equation,

$$\Delta G^\ominus = -RT\ln K_{eq},$$

we obtain,

$$\ln K_{eq} = -\frac{\Delta H^\ominus}{RT} + \frac{\Delta S^\ominus}{R}.$$

Differentiation of this expression with respect to the variable T yields the Van 't Hoff equation.

Provided that $\Delta H^\ominus$ and $\Delta S^\ominus$ are constant, the preceding equation gives $\ln K$ as a linear function of $1/T$ and hence is known as the *linear form* of the Van 't Hoff equation. Therefore, when the range in temperature is small enough that the standard enthalpy and entropy changes are essentially constant, a plot of the natural logarithm of the equilibrium constant versus the reciprocal temperature gives a straight line. The slope of the line may be multiplied by the gas constant R to obtain the standard enthalpy change of the reaction, and the intercept may be multiplied by R to obtain the standard entropy change.

Van 't Hoff Isotherm

The Gibbs free energy changes with the temperature and pressure of the thermodynamic system. The Van 't Hoff isotherm can be used to determine the Gibbs free energy for non-standard state reactions at a constant temperature:

$$\left(\frac{dG}{d\xi}\right)_{T,p} = \Delta_r G + RT\ln Q_r,$$

where $\Delta_r G$ is the Gibbs free energy for the reaction, and Q_r is the reaction quotient. When a reaction is at equilibrium, $Q_r = K_{eq}$. The Van 't Hoff isotherm can help estimate the equilibrium reaction shift. When $\Delta_r G < 0$, the reaction moves in the forward direction. When $\Delta_r G > 0$, the reaction moves in the backwards directions.

Van 't Hoff Plot

For a reversible reaction, the equilibrium constant can be measured at a variety of temperatures. This data can be plotted on a graph with ln K_{eq} on the y-axis and $1/T$ on the x axis. The data should have a linear relationship, the equation for which can be found by fitting the data using the linear form of the Van 't Hoff equation,

$$\ln K_{eq} = -\frac{\Delta H^{\ominus}}{RT} + \frac{\Delta S^{\ominus}}{R}$$

This graph is called the Van 't Hoff plot and is widely used to estimate the enthalpy and entropy of a chemical reaction. From this plot, $-\Delta H/R$ is the slope, and $\Delta S/R$ is the intercept of the linear fit.

By measuring the equilibrium constant, K_{eq}, at different temperatures, the Van 't Hoff plot can be used to assess a reaction when temperature changes. Knowing the slope and intercept from the Van 't Hoff plot, the enthalpy and entropy of a reaction can be easily obtained using:

$$\Delta H = -R \times \text{slope},$$
$$\Delta S = R \times \text{intercept}$$

The Van 't Hoff plot can be used to quickly determine the enthalpy of a chemical reaction both qualitatively and quantitatively. Change in enthalpy can be positive or negative, leading to two major forms of the Van 't Hoff plot.

Endothermic Reactions

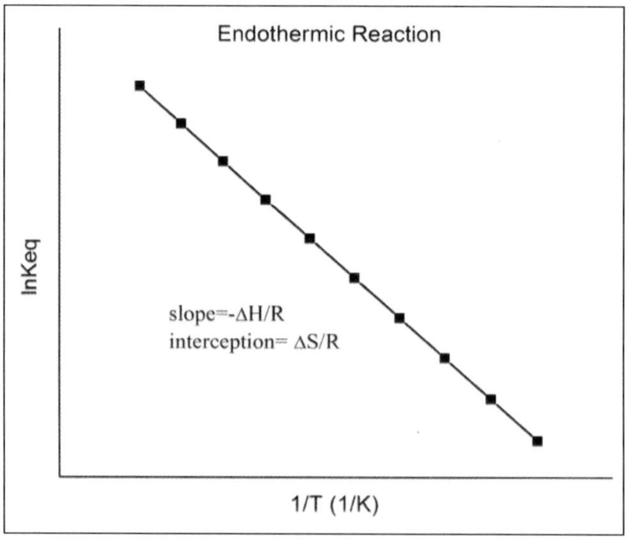

Van 't Hoff plot for an endothermic reaction.

For an endothermic reaction, heat is absorbed, making the net enthalpy change positive. Thus, according to the definition of the slope,

$$\text{slope} = -\frac{\Delta H}{R},$$

for an endothermic reaction, $\Delta H > 0$ (and the gas constant $R > 0$). So,

$$\text{slope} = -\frac{\Delta H}{R} < 0.$$

Thus, for an endothermic reaction, the Van ‘t Hoff plot should always have a negative slope.

Exothermic Reactions

For an exothermic reaction, heat is released, making the net enthalpy change negative. Thus, according to the definition of the slope,

$$\text{slope} = -\frac{\Delta H}{R},$$

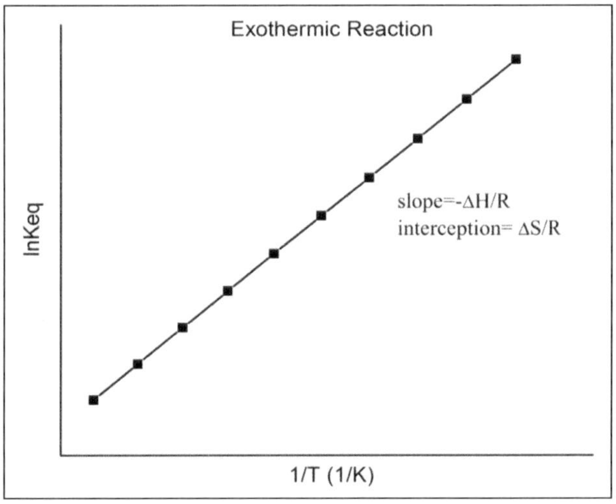

Van ‘t Hoff plot for an exothermic reaction.

from an exothermic reaction, $\Delta H < 0$. So,

$$\text{slope} = -\frac{\Delta H}{R} > 0.$$

Thus, for an exothermic reaction, the Van ‘t Hoff plot should always have a positive slope.

Error Propagation

Using the fact that $\Delta G^\ominus = -RT \ln K = \Delta H^\ominus - T\Delta S^\ominus$, it would appear that two measurements of K would suffice to be able to obtain a value of $\Delta H^\ominus$,

$$\Delta H^\ominus = R\frac{\ln K_1 - \ln K_2}{\frac{1}{T_2} - \frac{1}{T_1}},$$

where K_1 and K_2 are the equilibrium constant values obtained at temperatures T_1 and T_2 respectively. The precision of $\Delta H^\ominus$ values obtained in this way is highly dependent on the precision of the equilibrium constant values. A typical pair of temperatures might be 25 and 35 °C (298 and 308 K). For these temperatures,

$$\frac{1}{T_2}-\frac{1}{T_1}=\frac{1}{298}-\frac{1}{308}=1.1\times10^{-4}\ \text{K}.$$

Inserting this value into the expression for $\Delta H^{\ominus}$:

$$\Delta H^{\ominus} \approx 76\left(\ln K_1 - \ln K_2\right)\ \text{kJ/mol}.$$

Now, error propagation shows that the error on $\Delta H^{\ominus}$ will be about 76 kJ/mol times the error on ($\ln K_1 - \ln K_2$), or about 110 kJ/mol times the error on the ln *K* values. Assume for example that the error on each ln *K* is $\sigma \approx 0.05$, a small but reasonable value. The error on $\Delta H^{\ominus}$ will be about 5 kJ/mol. So, even though the individual stability constants were determined with good precision, the enthalpy calculated in this way is subject to a significant error.

The entropy will then be obtained from $\Delta S^{\ominus} = 1/T(\Delta H^{\ominus} + RT \ln K)$. In this expression the error on the second term is negligible compared to the error on the first term. The magnifying factor is then 76 kJ/mol ÷ 298 K, so for an error of 0.05 in the logarithms the error on $\Delta S^{\ominus}$ will be of the order of 17 J/(K mol).

When equilibrium constants are measured at three or more temperatures, values of $\Delta H^{\ominus}$ will be obtained by straight-line fitting. In this case the error on the standard enthalpy will be magnified to a somewhat lesser, but still substantial, extent.

Applications of the Van 't Hoff Plot

Van 't Hoff Analysis

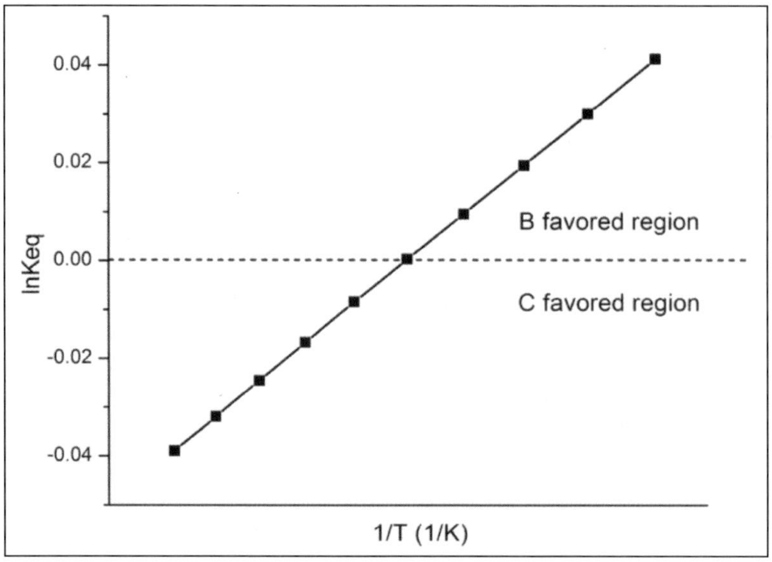

Van 't Hoff analysis.

In biological research, the Van 't Hoff plot is also called Van 't Hoff analysis. It is most effective in determining the favored product in a reaction.

Assume two products B and C form in a reaction:

$$a\,\mathrm{A} + d\,\mathrm{D} \rightarrow b\,\mathrm{B},$$

$$a\,\mathrm{A} + d\,\mathrm{D} \rightarrow c\,\mathrm{C}.$$

In this case, K_{eq} can be defined as ratio of B to C rather than the equilibrium constant.

When $B/C > 1$, B is the favored product, and the data on the Van 't Hoff plot will be in the positive region.

When $B/C < 1$, C is the favored product, and the data on the Van 't Hoff plot will be in the negative region.

Using this information, a Van 't Hoff analysis can help determine the most suitable temperature for a favored product.

In 2010, a Van 't Hoff analysis was used to determine whether water preferentially forms a hydrogen bond with the *C*-terminus or the *N*-terminus of the amino acid proline. The equilibrium constant for each reaction was found at a variety of temperatures, and a Van 't Hoff plot was created. This analysis showed that enthalpically, the water preferred to hydrogen bond to the *C*-terminus, but entropically it was more favorable to hydrogen bond with the *N*-terminus. Specifically, they found that *C*-terminus hydrogen bonding was favored by 4.2–6.4 kJ/mol. The *N*-terminus hydrogen bonding was favored by 31–43 J/(K mol).

This data alone could not conclude which site water will preferentially hydrogen-bond to, so additional experiments were used. It was determined that at lower temperatures, the enthalpically favored species, the water hydrogen-bonded to the *C*-terminus, was preferred. At higher temperatures, the entropically favored species, the water hydrogen-bonded to the *N*-terminus, was preferred.

Mechanistic Studies

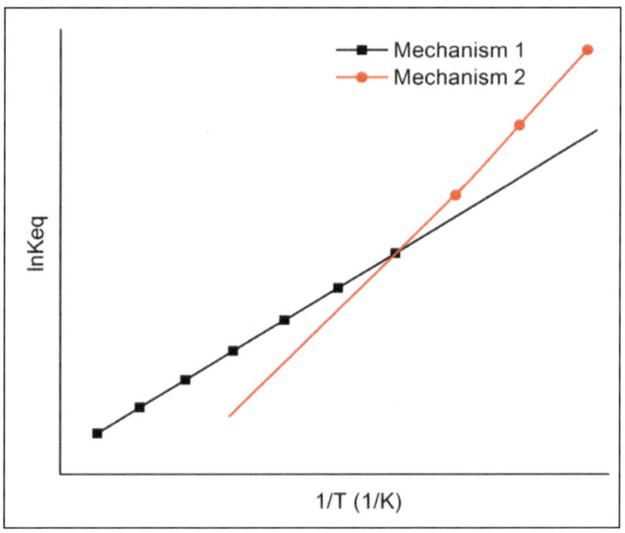

Van 't Hoff plot in mechanism study.

A chemical reaction may undergo different reaction mechanisms at different temperatures.

In this case, a Van 't Hoff plot with two or more linear fits may be exploited. Each linear fit has a different slope and intercept, which indicates different changes in enthalpy and entropy for each distinct mechanisms. The Van 't Hoff plot can be used to find the enthalpy and entropy change for each mechanism and the favored mechanism under different temperatures.

$$\Delta H_1 = -R \times \text{slope}_1, \quad \Delta S_1 = R \times \text{intercept}_1;$$
$$\Delta H_2 = -R \times \text{slope}_2, \quad \Delta S_2 = R \times \text{intercept}_2.$$

In the example figure, the reaction undergoes mechanism 1 at high temperature and mechanism 2 at low temperature.

Temperature Dependence

The Van 't Hoff plot is linear based on the assumption that the enthalpy and entropy are constant with temperature changes. However, in some cases the enthalpy and entropy do change dramatically with temperature. A first-order approximation is to assume that the two different reaction products have different heat capacities. Incorporating this assumption yields an additional term c/T^2 in the expression for the equilibrium constant as a function of temperature. A polynomial fit can then be used to analyze data that exhibits a non-constant standard enthalpy of reaction:

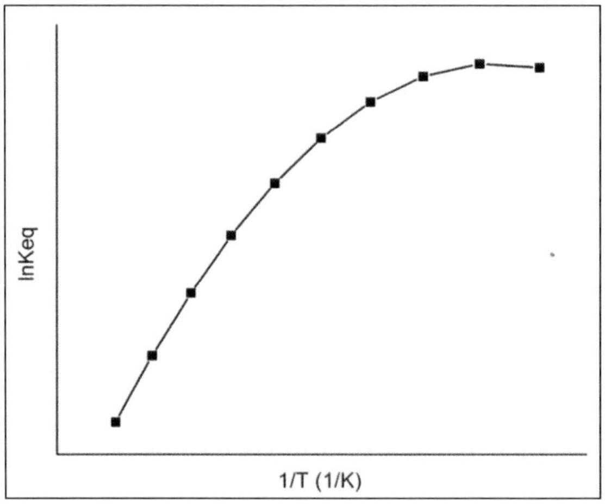

Temperature-dependent Van 't Hoff plot.

$$\ln K_{eq} = a + \frac{b}{T} + \frac{c}{T^2},$$

where,

$$\Delta H = -R\left(b + \frac{2c}{T}\right),$$

$$\Delta S = R\left(a - \frac{c}{T^2}\right).$$

Thus, the enthalpy and entropy of a reaction can still be determined at specific temperatures even when a temperature dependence exists.

Surfactant Self-assembly

The Van 't Hoff relation is particularly useful for the determination of the micellization enthalpy $\Delta H^{\ominus}_{m}$ of surfactants from the temperature dependence of the critical micelle concentration (CMC):

$$\frac{d}{dT}\ln \text{CMC} = \frac{\Delta H^{\ominus}_{\text{m}}}{RT^2}.$$

However, the relation loses its validity when the aggregation number is also temperature-dependent, and the following relation should be used instead,

$$RT^2\left(\frac{\partial}{\partial T}\ln \text{CMC}\right)_P = -\Delta H^{\ominus}_{\text{m}}(N) + T\left(\frac{\partial}{\partial N}(G_{N+1} - G_N)\right)_{T,P}\left(\frac{\partial N}{\partial T}\right)_P,$$

with G_{N+1} and G_N being the free energies of the surfactant in a micelle with aggregation number $N + 1$ and N respectively. This effect is particularly relevant for nonionic ethoxylated surfactants or polyoxypropylene–polyoxyethylene block copolymers (Poloxamers, Pluronics, Synperonics). The extended equation can be exploited for the extraction of aggregation numbers of self-assembled micelles from differential scanning calorimetric thermograms.

Le Chatelier's Principle

Le Chatelier's Principle is the principle when a stress is applied to a chemical system at equilibrium, the equilibrium will shift to relieve the stress. In other words, it can be used to predict the direction of a chemical reaction in response to a change in conditions of temperature, concentration, volume, or pressure. While Le Chatelier's principle can be used to predict the response to a change in equilibrium, it does not explain (at a molecular level), why the system responds as it does.

Chatalier's Principle or the Equilibrium Law

The principle is named for Henry Louis Le Chatelier. Le Chatelier and Karl Ferdinand Braun independently proposed the principle, which is also known as Chatelier's principle or the equilibrium law. The law may be stated:

> When a system at equilibrium is subjected to a change in temperature, volume, concentration, or pressure, the system readjusts to partially counter the effect of the change, resulting in a new equilibrium.

While chemical equations are typically written with reactants on the left, an arrow pointing from left to right, and products on the right, the reality is that a chemical reaction is at equilibrium. In other words, a reaction may proceed in both the forward and backward direction or be reversible. At equilibrium, both the forward and back reactions occur. One may proceed much more quickly than the other.

In addition to chemistry, the principle also applies, in slightly different forms, to the fields of pharmacology and economics.

Use of Le Chatelier's Principle in Chemistry

Concentration: An increase in the amount of reactants (their concentration) will shift the equilibrium to produce more products (product-favored). Increasing the number of products will shift the reaction to make more reactants (reactant-favored). Decreasing reactants favors reactants. Decreasing product favors products.

Temperature: Temperature may be added to a system either externally or as a result of the chemical reaction. If a chemical reaction is exothermic (ΔH is negative or heat is released), heat is considered a product of the reaction. If the reaction is endothermic (ΔH is positive or heat is absorbed), heat is considered a reactant. So, increasing or decreasing temperature can be considered the same as increasing or decreasing the concentration of reactants or products. In the temperature is increased, the heat of the system increases, causing the equilibrium to shift to the left (reactants). If the temperature is decreased, the equilibrium shifts to the right (products). In other words, the system compensates for the reduction in temperature by favoring the reaction that generates heat.

Pressure/Volume: Pressure and volume can change if one or more of the participants in a chemical reaction is a gas. Changing the partial pressure or volume of a gas acts the same as changing its concentration. If the volume of gas increases, pressure decreases (and vice versa). If the pressure or volume increase, the reaction shifts toward the side with lower pressure. If the pressure is increased or volume decreases, equilibrium shifts toward the higher pressure side of the equation. Note, however, that adding an inert gas (e.g., argon or neon) increases the overall pressure of the system, yet does not change the partial pressure of the reactants or products, so no equilibrium shift occurs.

References

- Hefter, G.T.; Tomkins, R. P. T., eds. (2003). The Experimental Determination of Solubilities. Wiley. ISBN 0-471-49708-8
- Chemical-equilibrium, science: britannica.com, Retrieved 18 April, 2019
- Mendham, J.; Denney, R. C.; Barnes, J. D.; Thomas, M. J. K. (2000), Vogel's Quantitative Chemical Analysis (6th ed.), New York: Prentice Hall, ISBN 0-582-22628-7
- Law-of-chemical-equilibrium-constant, equilibrium, chemistry, guides: toppr.com, Retrieved 13 July, 2019
- Project: Ionic Strength Corrections for Stability Constants". IUPAC. Archived from the original on 2008-10-29. Retrieved 2008-11-23
- Atkins, P. W.; De Paula, J. (2006). "Section 7.4: The Response of Equilibria to Temperature". Physical Chemistry. Oxford University Press. ISBN 0-19-870072-5
- Project: Ionic Strength Corrections for Stability Constants". International Union of Pure and Applied Chemistry. Retrieved 2019-03-28
- Definition-of-le-chateliers-principle-605297: thoughtco.com, Retrieved 27 May, 2019

PERMISSIONS

All chapters in this book are published with permission under the Creative Commons Attribution Share Alike License or equivalent. Every chapter published in this book has been scrutinized by our experts. Their significance has been extensively debated. The topics covered herein carry significant information for a comprehensive understanding. They may even be implemented as practical applications or may be referred to as a beginning point for further studies.

We would like to thank the editorial team for lending their expertise to make the book truly unique. They have played a crucial role in the development of this book. Without their invaluable contributions this book wouldn't have been possible. They have made vital efforts to compile up to date information on the varied aspects of this subject to make this book a valuable addition to the collection of many professionals and students.

This book was conceptualized with the vision of imparting up-to-date and integrated information in this field. To ensure the same, a matchless editorial board was set up. Every individual on the board went through rigorous rounds of assessment to prove their worth. After which they invested a large part of their time researching and compiling the most relevant data for our readers.

The editorial board has been involved in producing this book since its inception. They have spent rigorous hours researching and exploring the diverse topics which have resulted in the successful publishing of this book. They have passed on their knowledge of decades through this book. To expedite this challenging task, the publisher supported the team at every step. A small team of assistant editors was also appointed to further simplify the editing procedure and attain best results for the readers.

Apart from the editorial board, the designing team has also invested a significant amount of their time in understanding the subject and creating the most relevant covers. They scrutinized every image to scout for the most suitable representation of the subject and create an appropriate cover for the book.

The publishing team has been an ardent support to the editorial, designing and production team. Their endless efforts to recruit the best for this project, has resulted in the accomplishment of this book. They are a veteran in the field of academics and their pool of knowledge is as vast as their experience in printing. Their expertise and guidance has proved useful at every step. Their uncompromising quality standards have made this book an exceptional effort. Their encouragement from time to time has been an inspiration for everyone.

The publisher and the editorial board hope that this book will prove to be a valuable piece of knowledge for students, practitioners and scholars across the globe.

INDEX